Geological Storage of CO_2

Geological Storage of CO_2
Modeling Approaches for Large-Scale Simulation

Jan M. Nordbotten
Department of Mathematics
University of Bergen

Michael A. Celia
Department of Civil and Environmental Engineering
Princeton University

WILEY

A John Wiley & Sons, Inc., Publication

Published by John Wiley & Sons, Inc., Hoboken, New Jersey

Published simultaneously in Canada

Library of Congress Cataloging-in-Publication Data:

Nordbotten, Jan Martin.
Geological storage of CO_2 : modeling approaches for large-scale simulation / Jan M. Nordbotten, Michael A. Celia.
p. cm.
Includes index.
ISBN 978-0-470-88946-6 (hardback)
1. Geological carbon sequestration—Mathematical models. I. Celia, Michael Anthony. II. Title.
QE516.C37C45 2012
628.5′3—dc23

2011019942

Printed in the United States of America

oBook ISBN: 9781118137086
ePDF ISBN: 9781118137055
ePub ISBN: 9781118137079
MOBI ISBN: 9781118137062

10 9 8 7 6 5 4 3 2 1

Contents

Prologue

Injection of carbon dioxide into subsurface formations dates back to the early 1970s. At that time, there was no consideration of the environmental benefit; rather, the purpose was to provide pressure and solubility conditions which would be advantageous for increased oil production. More than a decade later, in the 1980s, injection for the purpose of emission avoidance was first considered. This, together with favorable legislation, led to the first pure CO_2 storage project, off the coast of western Norway in 1996. Since then, carbon storage has gained increasing attention as a carbon mitigation option from academia, industry, environmental organizations, and regulators.

Despite this relatively long history, and the significant current interest, no textbooks exist that cover the fundamentals of carbon storage. As such, students of the subject are forced to assimilate knowledge from a broad range of sources. This situation is unfortunate, in that it hinders the development of dedicated courses for teaching the science and technology of carbon storage, and consequently makes the subject less accessible in general. We hope that this book will begin to remedy this situation.

As the first textbook on the subject of carbon storage, we hope our audience will be broad. However, this is not a general book on carbon mitigation, or even on the overall technology of carbon capture and storage. Rather, we have focused the book on basic concepts needed to understand subsurface storage of CO_2, with a focus on mathematical models used to describe storage operations. We have begun from the basic concepts of flow in porous media, and expanded the discussion to include fairly involved mathematical developments, especially in the later chapters. We have attempted to make the early material accessible to a wide audience, with the successive chapters more oriented toward graduate-level students in the fields of civil and environmental engineering, subsurface physics, petroleum engineering, and applied mathematics. At the same time, we hope that experienced professionals from related fields will find in this book a concise introduction to the particular challenges that separates carbon storage from other subsurface problems. Finally, the regulator will hopefully find in this book descriptions of the key physical processes, of which knowledge is needed in the design and enforcement of meaningful regulation.

This is not a long book. Nevertheless, we attempt to take the reader all the way from the basic concepts of fluid flow in porous formations to considerations involving large-scale simulation. This is a steep journey; however, we are confident that the text will allow the dedicated reader to follow us. The first chapter serves as the motivation for this journey. There, we provide arguments for the role of carbon storage in the global portfolio of carbon emission reduction strategies. We also review the basic features of geology that makes the idea of carbon storage plausible.

Our second and third chapters cover the laws that govern fluid motion in the porous materials that constitute the subsurface. In Chapter 2, we consider flow of a single fluid. In addition to the laws of motion, we review the basic solution strategies for the governing equations, together with the most common simplifications. This approach is mirrored in Chapter 3, where we focus on the flow of multiple fluids in the pore space, with a general focus on two-fluid systems involving supercritical CO_2 and brine. The classical theory in both these chapters is presented while emphasizing the particular aspects most relevant for carbon storage.

Chapter 4 considers changes of scale. In particular, it addresses the fact that our governing equations are developed and parameterized in the laboratory, while the enormous volumetric needs of carbon storage force us to answer questions on scales spanning many kilometers horizontally and decades in time. Bridging the gap between the laboratory and field is therefore a crucial aspect. In this chapter we present a systematic approach to allow equations to be written at large space and time scales while still retaining essential information from smaller scales.

With large-scale equations in hand, the two last chapters discuss solution approaches. Analytical and numerical solution methods for flows contained within a single aquifer are covered in Chapter 5, including practical calculations applied to real-world data. In Chapter 6, additional issues related to the interaction with other aquifers in the vertical sequence, in terms of pressure propagation and fluid migration, are covered. We conclude Chapter 6 with a set of practical calculations, involving leakage across different formations in a vertical sequence, using as realistic data as possible. This allows us to emphasize the important practical quantities that can be calculated from the governing models.

In sum, this book contains what we see as the essentials of carbon storage modeling, with a focus on how systems and solutions can be simplified to provide useful solutions to practical questions. It is the book we wanted to write. We can only hope that it is also the book that the audience wants to read.

We have greatly enjoyed writing this book, and we have also enjoyed the great fortune to be located at two outstanding educational institutions where we have many colleagues interested in the carbon problem. At Princeton, the support of the Carbon Mitigation Initiative (CMI) has been invaluable, and the authors are especially thankful to the co-directors of CMI, Steve Pacala and Rob Socolow. At the University of Bergen, the Center of Integrated Petroleum Research has provided both assistance and a wealth of interested colleagues, most notably our long-time colleagues Magne Espedal and Helge Dahle. We have also been fortunate to work with many other terrific colleagues and students, both from our home institutions and from many other institutions around the world. Very helpful reviews of different versions of the text were provided by Mark Person from the New Mexico Institute of Technology and Mining, Karl Bandilla from Princeton University, Knut-Andreas Lie from SINTEF ICT, and especially Quanlin Zhou from Lawrence Berkeley National Laboratory. We are very much indebted to these reviewers. Finally, much of the writing was done during extended visits to the Ecole Polytechnique Federale de

Lausanne (EPFL), the University of Utrecht, and the University of Barcelona. We thank our very gracious hosts at these institutions—Andrea Rinaldo, Majid Hassanizadeh, and Jesus Carrera—for their help and their understanding as we "hid out" to concentrate on our writing. The result is this book, which we very much hope will help in the efforts to solve the carbon problem.

Bergen and Princeton
August 2011

Jan M. Nordbotten
Michael A. Celia

Chapter 1

The Carbon Problem

The "carbon problem" refers to the ongoing increase in atmospheric concentrations of the greenhouse gas carbon dioxide (CO_2) observed over the last two centuries. This increase is being driven almost entirely by anthropogenic emissions, with most of the emissions associated with combustion of fossil fuels. If humankind decides, at some point, to reduce significantly the anthropogenic emission rate, new or different technologies will almost certainly play a central role. In this book, we are interested in one specific technology: carbon capture and storage (CCS), wherein the CO_2 produced from the use of fossil fuels is captured at large stationary sources like power plants and is stored somewhere other than the atmosphere. We are specifically interested in the storage part of this operation, and even more specifically, in geological storage, where the captured CO_2 is injected into appropriate geological formations deep underground. Proper analysis of the operations and possible consequences of this kind of injection requires careful mathematical and computational models to predict the system behavior. We focus on such models in this book.

1.1 BACKGROUND

The concentration of CO_2 in the atmosphere is naturally dynamic. Figure 1.1 shows the so-called Keeling curve, named after Charles Keeling, who initiated a program of ongoing measurements of atmospheric CO_2 in the 1950s. These data show annual cycles of variability superimposed on a monotonic increase over the half century of measurements. Atmospheric concentration of CO_2 in the late 1950s was around 315 parts per million (ppm), while today's concentration has grown to about 390 ppm. To put these numbers into historical context, consider the data shown in Figure 1.2. There the Keeling data, measured at Mauna Loa, Hawaii, are combined with data from ice cores to show atmospheric concentration of CO_2 over the last 1000 years. These data show a stable atmospheric concentration of about 280 ppm, which is the concentration to which the atmosphere stabilized at the end of the last ice age. The increase above 280 ppm began with the industrial revolution and has accelerated continuously to the present day. The range of values shown in Figure 1.2, between

Geological Storage of CO_2: Modeling Approaches for Large-Scale Simulation, First Edition.
Jan M. Nordbotten and Michael A. Celia.

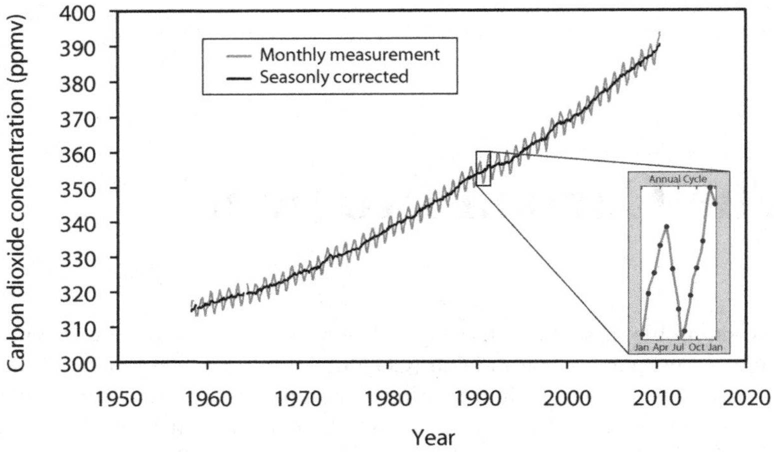

Figure 1.1 Atmospheric carbon dioxide as a function of time, measured at Mauna Loa (modified from wikipedia.org/wiki/Keeling_curve).

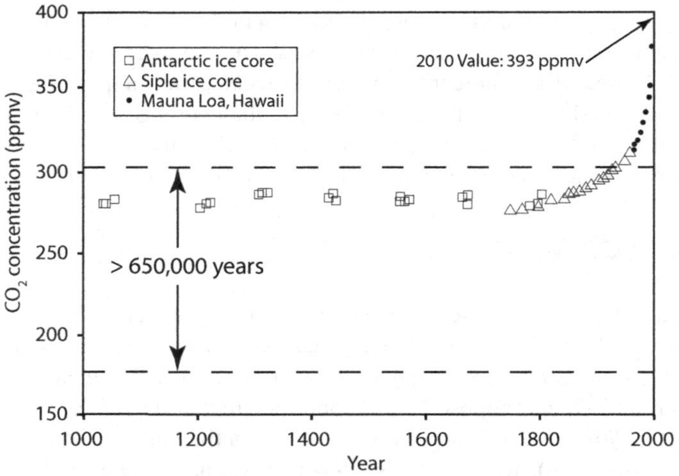

Figure 1.2 Atmospheric carbon dioxide as a function of time, over the past 1000 years. Also shown is the range of CO_2 concentrations measured over the past 650,000 years (modified from http://www.britannica.com/eb/art-69345/Carbon-dioxide-concentrations-in-Earths-atmosphere-plotted-over-the-past).

a low of about 170 ppm and a high of about 300 ppm, indicates the maximum and minimum values of atmospheric CO_2 concentration seen in ice core data over the last 650,000 years. In those ice core records, clear 100,000-year cycles of glacial–interglacial periods can be seen, with corresponding maximum and minimum values of atmospheric CO_2. From these data, we conclude that the current concentration is about 100 ppm above the "natural" equilibrium associated with the current interglacial period. We also conclude that the current value of 390 ppm is larger, by about

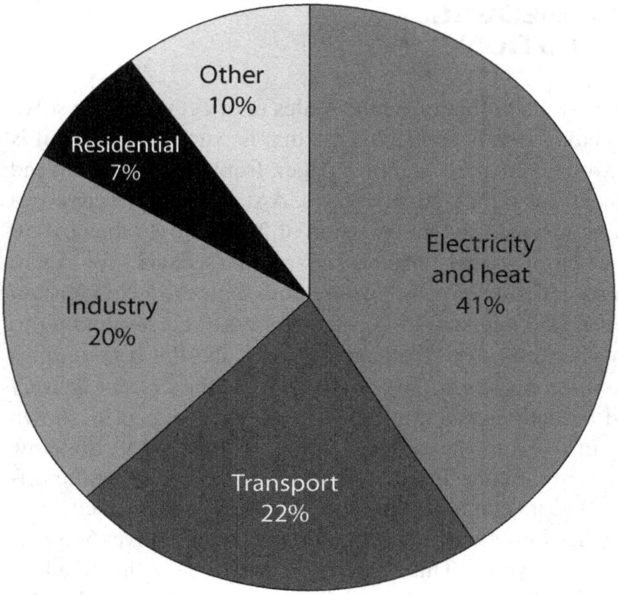

Figure 1.3 Fraction of total fossil fuel emissions by sector. Data from the International Energy Agency (IEA 2010).

30%, than the highest value seen in at least the last 650,000 years. As such, we humans are collectively performing an interesting global-scale experiment to see how the earth system will respond to significant increases of an important greenhouse gas. The consensus expectation is that these increases will lead to dangerous climate change unless they are reduced or reversed.

In order to understand the problem, it is helpful to identify the specific sources of anthropogenic CO_2 associated with combustion of fossil fuels. Figure 1.3 shows global estimates for the different sources of CO_2 emissions, indicating the fraction of total fossil fuel-related emissions coming from each major sector. The dominant source of CO_2 emissions is electricity generation, accounting for approximately 40% of emissions, followed by transportation at slightly more than 20%. A recent estimate for total annual anthropogenic emissions (for calendar year 2008) is between 8 and 9 gigatonnes (Gt) of carbon (where 1 Gt C = 1×10^{15} g of carbon = 1×10^{9} metric tones of carbon). The molecular weight of carbon is 12, while the weight of a CO_2 molecule is 44; therefore, the conversion factor from carbon to CO_2 is 3.67, which means that the global annual anthropogenic emission rate measured in mass of CO_2 is about 30 Gt CO_2/year. In the remainder of this section we will use 8 Gt C/year as the estimated current emission rate.

Given this profile of emissions, it seems logical that any successful strategy for carbon mitigation will involve decarbonization of electricity generation coupled with associated strategies that may include the use of decarbonized electricity in the remaining sectors. For example, one can consider electrification of the transportation sector and modified designs for both residential and commercial buildings to take advantage of carbon-free electricity. Overall, development of effective solutions to the carbon problem constitutes a grand challenge for the early 21st century.

1.2 STABILIZATION WEDGES AND TECHNICAL SOLUTIONS

To understand the overall size of the problem and the scales of effort needed to solve it, specific units of measure have been introduced. One that is particularly useful is the so-called stabilization wedge. In their seminal paper from 2004, Pacala and Socolow introduced the concept of stabilization wedges. A stabilization wedge is a unit of measure corresponding to avoidance of emission of 25 Gt C over the next 50 years. The concept arises when a typical "business-as-usual" scenario for future emissions is compared to a so-called "flat path," where emissions are held constant at their current rate for the next 50 years. With business as usual corresponding to doubling of the current emission rate over the next 50 years, the flat path implies that instead of increasing emission rates to 16 Gt C/year from the current 8 Gt C/year, the rate would instead be held constant at 8 Gt C/year. The concept is shown in Figure 1.4, which is modified from the original figure of Pacala and Socolow (2004), which used the then-current emission rate of 7 Gt/year. We see that the difference of 8 Gt C/year after 50 years can be broken into 8 "slices," where each slice, or "wedge," corresponds to emissions avoidance that increases linearly over 50 years to a value of 1 Gt C/year after 50 years. This "slice," or "wedge," is indicated on Figure 1.4. Among many other things, Pacala and Socolow (2004) considered a variety of existing technologies and estimated the effort needed to implement those technologies in order to achieve one wedge. A partial list of available technologies, and the associated effort to achieve one wedge worth of emissions avoidance, is given in Table 1.1 (this has also been updated from the original 2004 publication).

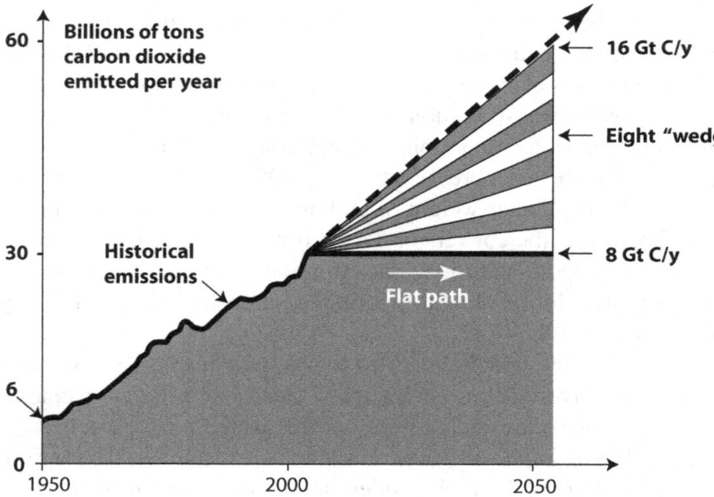

Figure 1.4 Stabilization wedges, modified from the original presented in Pacala and Socolow (2004), which used a "current" emission rate of 7 Gt C/year.

Table 1.1 Examples of Technologies and Associated Efforts to Achieve One Wedge (Updated from Pacala and Socolow 2004)

Technology	Effort required to achieve one wedge
Automobile efficiency	Increase fuel efficiency from 30 mpg to 60 mpg for 2 billion cars (currently there are fewer than 1 billion cars worldwide).
Nuclear power	Add twice the currently installed nuclear capacity.
Solar power	Add 350 times the currently installed solar power capacity.
Wind power	Add 15 times the currently installed capacity of wind-generated electricity.
Carbon capture and storage	Install carbon capture and storage at 800 large-scale coal-fired power plants.

Notable technologies include nuclear power, which requires addition of twice the current global installed capacity to achieve one wedge; solar photovoltaics, which require about 350 times the global installed capacity (as of 2008) to achieve one wedge; automobile efficiency, which requires 2 billion cars to increase fuel efficiency from 30 miles per gallon (mpg) to 60 mpg (there are currently about 650 million cars worldwide); and installation of a technology called CCS at about 800 large-scale coal-fired power plants. Based on these calculations, Pacala and Socolow (2004) made three important points. The first is that several current technologies exist that can produce at least one wedge of emissions avoidance. The second is that each technology requires an enormous effort to reach one wedge. And the third is that none of the technologies is capable of producing the entire eight wedges needed to achieve the flat path for future emissions. As such, a portfolio of technologies must be used if the carbon problem is to be solved.

When considering energy portfolios of the countries that are the largest emitters, coal plays a central role. Abundant and cheap coal can be found in both the United States and China, which means that coal-fired electricity is likely to be a significant part of the energy portfolio in each country for at least the next several decades. This makes decarbonization of the power grid a problem of use of coal without the CO_2 emissions. The only currently available technology to achieve this outcome is CCS. To see the potential impact of CCS technology, we can consider that there are currently about 2100 large coal-fired power plants worldwide. Furthermore, the current rate of construction of new plants in China alone is widely estimated at between one and two plants per week. Both of these numbers imply that a full-scale implementation of CCS coupled with coal has the possibility to contribute several wedges to the overall solution. And, as the Massachusetts Institute of Technology (MIT 2007) report on the future of coal concluded in 2007, "carbon capture and sequestration (CCS) is the critical enabling technology that would reduce CO_2 emissions significantly while also allowing coal to meet the world's pressing energy needs."

1.3 CCS

As a possible large-scale mitigation strategy for the atmospheric carbon problem, CCS has emerged as a serious option. The concept of CCS is simple: capture the CO_2 produced when the chemical energy in fossil fuels is converted to electrical energy, and sequester the captured carbon somewhere other than the atmosphere. The most likely location for large-scale sequestration is in deep geological formations, where the CO_2 would be injected into formations that are (1) sufficiently permeable to accept large quantities of CO_2 and (2) overlain by very low-permeability formations that will keep the injected buoyant CO_2 in place. Deep sedimentary basins are the likely target for large-scale CO_2 injections, using some combination of depleted oil and gas reservoirs (with or without enhanced oil recovery), unminable coal seams, and deep saline aquifers (see Figure 1.5). The recent Intergovernmental Panel on Climate Change (IPCC)'s special report on CCS provides many details about the overall concept and the technological approaches for its possible implementation. The report presents data showing that overall global storage capacity, in terms of total amounts of CO_2 that can be stored underground, appears to be more than adequate, with the largest capacity associated with deep saline aquifers. Figure 1.6, taken from the IPCC report, identifies promising areas for CO_2 storage across the globe.

The concept of CCS involves capture of the CO_2 prior to release to the atmosphere and storage (or sequestration) away from the atmosphere. In principle, technology exists to achieve both of these steps. Capture can be achieved through separation of the CO_2 from the dilute gaseous waste stream of traditional power plants. Or it can be achieved through the use of different designs to process the fuel and produce the electricity, for example, splitting the hydrogen and carbon in the

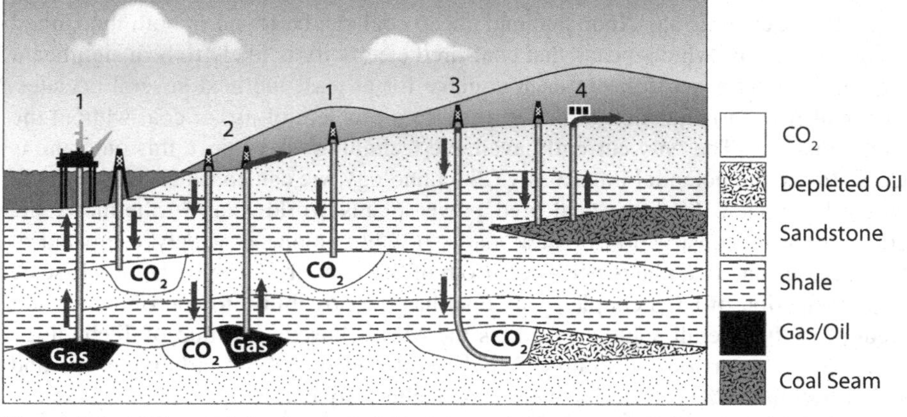

Figure 1.5 Different kinds of geological formations suitable for geological storage of CO_2 (image adapted from IPCC 2005).

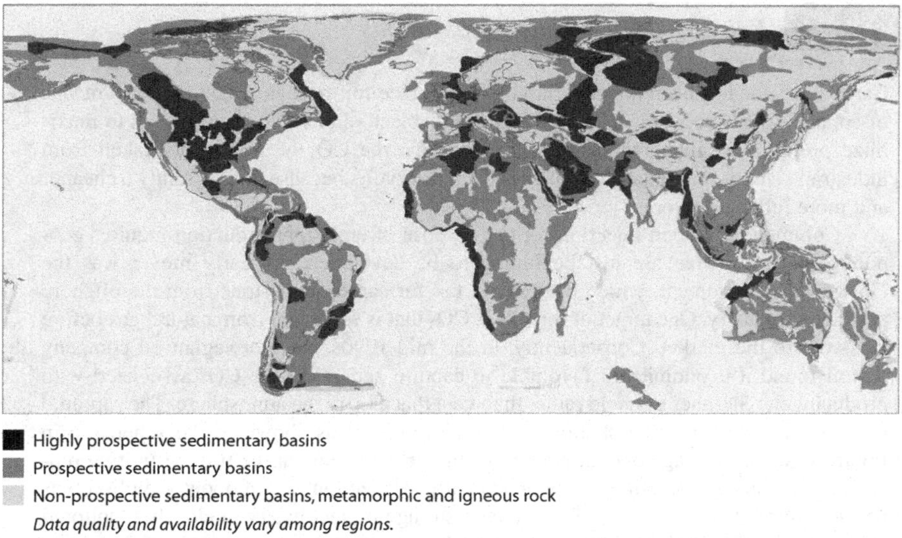

Figure 1.6 Prospective areas for geological storage of CO_2 (from IPCC 2005 report).

fossil fuel hydrocarbon early in the process and using the hydrogen as the energy carrier while producing an essentially pure stream of CO_2. This latter concept is the basis of the so-called integrated gasification combined cycle (IGCC) designs. The limiting factor in carbon capture at this time is cost, with 80–90% of the cost of CCS being associated with the capture process. While estimates still vary significantly, if a traditional coal-fired power plant were replaced by a CCS power plant, the added cost to consumers would be about 5 cents (U.S.)/kWh. Recognizing that implementation of large-scale CCS operations on many power plants will take several decades to implement, the annual cost increases would be a small fraction of 1 cent/kWh.

Once the CO_2 has been captured, it must be kept out of the atmosphere. In geological storage, the captured CO_2 is compressed and injected into deep geological formations. Almost all strategies considered to date involve injection at sufficient depth so that both the pressure and temperature exceed the critical point of CO_2: 7.4 mega-Pascal (MPa) and 31.1°Celsius. Supercritical CO_2 is significantly denser than gaseous CO_2. Therefore, the same amount of mass occupies a much smaller volume and has a concomitantly weaker buoyant drive. For typical geothermal gradients, CO_2 is supercritical at depths below about 800 m.

One of the main advantages of CCS is that all of the needed technology already exists. Therefore, large-scale implementation is not limited by the need for new technological developments, although better capture technologies would improve efficiencies and lower cost, and new technologies may be needed for longer-term monitoring of the injection system. In addition, the oil industry and the waste disposal industry have extensive experience with injection of both gases and liquids

BOX 1.1 *Existing CO_2 Injection Operations*

Carbon dioxide has been injected into the subsurface for more than 25 years as a means of enhancing oil recovery (EOR). Because the objective of EOR operations is to maximize profit rather than reduce carbon emissions, the CO_2 is usually not taken from industrial sources, but rather from natural CO_2 reservoirs because it is currently a cheaper and more reliable option.

Commercial carbon injection for the purpose of emissions reduction requires economic incentives that are just beginning to be developed. An early mover was the Norwegian government, which instituted a tax for carbon emissions from the offshore petroleum industry. One target of this tax is CO_2 that is separated from natural gas before it is sold to the market. Consequently, in the mid-1990s, the Norwegian oil company Statoil found it economically favorable to capture and inject the CO_2 associated with producing the Sleipner gas field rather than vent that CO_2 to the atmosphere. The captured CO_2 was injected into the shallower Utsira formation, as shown in the schematic in Figure 1.A. This storage operation came online in 1996, and at the time of writing represents 14 years of operational experience with an injection rate of about 1 million tons per year through a single well. The Sleipner storage project has been closely monitored with a series of seismic measurements, which have provided large amounts of data on CO_2 migration within the formation.

A second commercial-scale ongoing storage operation is being conducted by BP and their partners at their In Salah gas field in Algeria. Since 2004, close to one million tons have been injected annually into a water-filled portion of the formation from which gas is produced.

It is worth noting that both the Sleipner and In Salah operations are situated in geographically challenging locations: Sleipner is in the middle of the North Sea, while In Salah is in the Sahara Desert. Still, the technological challenges associated with injec-

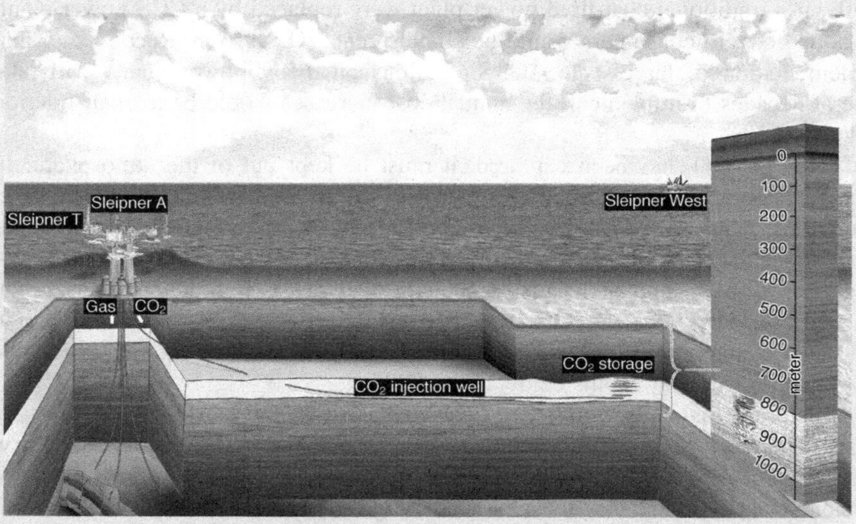

Figure 1.A Schematic of the Sleipner injection (courtesy of Statoil ASA).

tion have been overcome. The major challenge is therefore not in the technical implementation, but in our understanding of the subsurface system and our ability to predict the fate of injected CO_2, especially if CCS is implemented at a scale that is significant (at least one wedge) relative to solving the carbon problem.

into deep subsurface formations. This includes injection of significant quantities of liquid hazardous wastes in the United States and so-called "acid gas" (a mixture of CO_2 and hydrogen sulfide [H_2S]) in Canada. However, there are other issues that have slowed the implementation of large-scale CCS projects. These include the lack of a regulatory framework within which large-scale injection operations would be permitted, the lack of an economic system where the cost of emitting CO_2 to the atmosphere is internalized within an individual operator such as a power plant, and the lack of international agreements that would lead to effective global solutions.

When both regulations and economic credit systems are considered, a common question focuses on the fate of the injected CO_2, and more generally, the overall effects of large-scale injection on the subsurface system. If a significant amount of the injected CO_2 leaks out of the formation into which it is injected, then the operation may be viewed as a failure from an environmental, regulatory, and economic–credit perspective. A broader analysis of the overall system shows that leakage of either the injected CO_2, or of brine that is displaced by the injected CO_2, is the major issue associated with geological storage of CO_2. Therefore, a comprehensive analysis of the overall subsurface system is required, including the movement and ultimate fate of the CO_2 and other subsurface fluids affected by the injection operation.

The complete physical description of the subsurface system involves the physics of multiphase fluid flow in porous media, phase behavior of the fluids involved, heat and energy transport, geomechanics, and strongly nonlinear geochemical reactions. When all of these processes are included, the problem can become very complex, and different kinds of numerical and computational challenges arise. Engineering judgments about when certain simplifying assumptions can be made, their implications, and their limitations, lead to many interesting questions about underlying physics, the nature of mathematical (and computational) modeling, and the overall engineering and scientific approaches that are most appropriate for a given set of problems. In order to make these kinds of decisions about physics and mathematics, an understanding of the subsurface environment is necessary. We now present an introduction to the fascinating environment of the earth's subsurface.

1.4 THE SUBSURFACE SYSTEM

To understand geological storage and develop mathematical models to describe the movement of fluids in the subsurface, including injected CO_2, we need to have some understanding of the subsurface environment. If we were to start at the land surface and begin to move downward, we would first encounter soil, which is familiar to everyone. Soil is composed of different kinds of solid particles, collectively forming the solid part of the soil. These solid particles are not bound together, but rather are loose, and are referred to as *unconsolidated*. Between the solid particles are spaces filled with one or more fluids; for soils, these fluids are usually air and water. The space filled with fluids is referred to as the *pore space* or the *void space*. Our interest is primarily in the movement of fluids within the pore space.

As we continue further downward in the shallow soil zone, the mineral composition of the solid particles will typically change, with less organic material at greater depths. At some point, we will begin to notice that more of the pore space is filled with water, and eventually, we will reach a zone where all of the pores are filled with water. This is referred to as the *saturated zone*, while the zone above it, in which water and air coexist, is the *unsaturated zone*. As we proceed deeper, we may encounter layers of different kinds of materials, indicative of the particular geological history of the area. In areas of sediment deposition, these layers may be a few meters to tens of meters to perhaps hundreds of meters thick. In coastal areas like the coastal plain of New Jersey, USA, we can see clear layered structures with the layering alternating between layers of fine-grained, clay-rich materials and layers of coarser-grained sand deposits. Figure 1.7 shows a cross section of the New Jersey coastal plain, extending from Philadelphia, PA to Atlantic City, NJ. Two comments are of interest at this point. First, sandy materials tend to be much more permeable than clays. This is because the pores in coarse-grained sands are much larger than those in clays, and the permeability of the material depends strongly on the pore sizes. The higher permeability means that water can flow much more easily through the sand formations than through the clays. Water is often extracted from these kinds of sandy formations through pumping wells, thereby using so-called *groundwater*

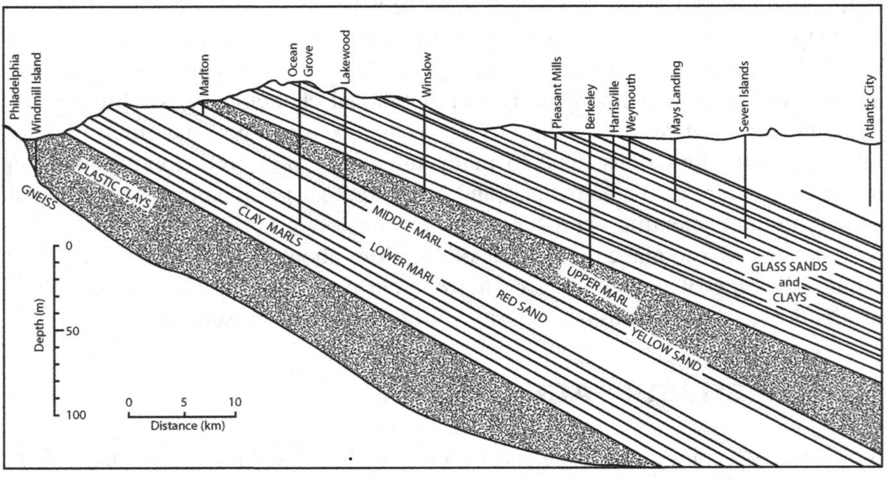

Figure 1.7 Sedimentary structure shown for a cross section of the New Jersey coastal plain (modified from Epstein 1987).

as a water supply. Groundwater supplies several billion people worldwide with water and is a vital resource. The second point to be made is that in Figure 1.7, the formations appear to have a significant slope toward the Atlantic Ocean. It is important to realize that this is a distortion of the true situation, because the scale on the vertical axis has been expanded, or exaggerated, by a factor of about 50 relative to the horizontal scale. If these were plotted with equal scales in the horizontal and vertical, the formations would be seen more properly as very thin, almost flat structures. It is a good idea to remember this vertical exaggeration and to interpret various images accordingly. It is also important to remember that the actual structures do indeed have much greater horizontal extent than vertical, and they commonly have slopes less than 1%.

Formations that contain freshwater and have permeability high enough to yield water in significant quantities for human use are referred to as *aquifers*. The lower permeability layers, like the clay layers in Figure 1.7, are often referred to as *confining layers* or *aquitards*. Aquitards do not allow water extraction in significant quantities, but they have enough permeability to allow for slow leakage of water across the formation, thereby providing hydraulic connection between aquifers immediately above and below the confining layer.

As we move deeper into the subsurface, we will often see similar layered structures. At depth, the rock tends to be *consolidated* rather than unconsolidated, but patterns of relatively permeable formations alternating with layers of very low permeability are often seen. A good example is the Alberta Basin, whose general vertical structure is characterized by alternating permeable (aquifer) and relatively impermeable (aquitard) formations. In deeper zones, the low-permeability layers are often referred to as *caprock formations*. Also, within the deeper formations, we encounter elevated pressures and temperatures, and the water tends to have high levels of dissolved solids, usually well above the level of salt found in seawater. Salt concentrations in the deeper formations across the Alberta Basin range from less than 50,000 ppm to more than 300,000 ppm. This compares to seawater at about 35,000 ppm. Water with high salt content is referred to as brine and is not considered suitable for either human consumption or agricultural use without significant desalination. Of course, the shallow zones continue to be filled with freshwater, and groundwater is an important resource in the Province of Alberta.

Thick sedimentary sequences like the Alberta Basin often contain oil and gas reservoirs. In Alberta, these producing reservoirs typically correspond to formations that would be called "aquifers," based on their flow properties. Existence of a mature oil and natural gas industry in Alberta has advantages for CO_2 storage operations because the properties of the subsurface, such as porosity and permeability, have been studied and characterized in detail, especially around the oil and gas fields. Furthermore, the associated infrastructure for oil and gas extraction operations might be used profitably for CO_2 injection projects. However, the development of the oil and gas fields has resulted in hundreds of thousands of wells having been drilled in the Alberta Basin. Many of these wells perforate tight confining units or seals (i.e., caprock formations), which have held hydrocarbons within underlying reservoirs over geological time scales. Without well perforations, these caprock

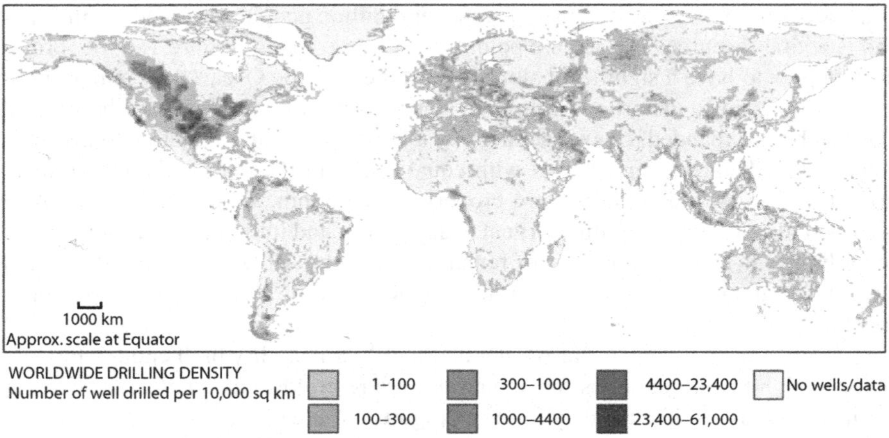

Figure 1.8 Density of wells (number per area) drilled across the world (from IPCC 2005).

formations are unlikely to allow CO_2 to leak should the CO_2 be injected into the underlying formation. However, the wells that have been drilled through the caprock formations might compromise the seal integrity because they provide potential leakage pathways for a buoyant fluid like CO_2. Leakage along wells is especially of concern in regions with a rich history of oil and gas exploration and production. This is apparent in Figure 1.8, which shows spatial densities of oil and gas wells worldwide, with North America in particular showing high densities of wells extending over significant distances.

The general approach for geological storage of CO_2 is to identify formations in the deep subsurface that have sufficiently high permeability to allow reasonable quantities of CO_2 to be injected. These are the formations corresponding to "aquifers" or "productive hydrocarbon reservoirs." The injection formation should be overlain by a low-permeability aquitard, or caprock formation, that has sufficiently low permeability to keep the CO_2 from migrating upward out of the injection formation. If the injection is into a deep brine-filled aquifer, then the CO_2 will displace the brine out of the pore space, so that CO_2 will now occupy a significant fraction of the pore space around the injection well. As the injection proceeds, the region occupied by CO_2 expands, and more brine is pushed outward away from the well. Because the CO_2 is significantly less dense than the resident brine, the injected CO_2 tends to migrate upward by buoyancy as it spreads away from the well, forming a characteristic displacement pattern within the formation. Of course, if wells have perforated the caprock, then some CO_2 and/or brine might leak along one or more of the wellbores, forming the more complex pattern of flow and leakage illustrated schematically in Figure 1.9. In addition to mass movement involving CO_2 and brine, the imposed pressure increase at the injection well also propagates into the formation, serving to drive the flow of both the CO_2 and the brine. How far this pressure pulse extends is also of interest, because it provides a way to define the

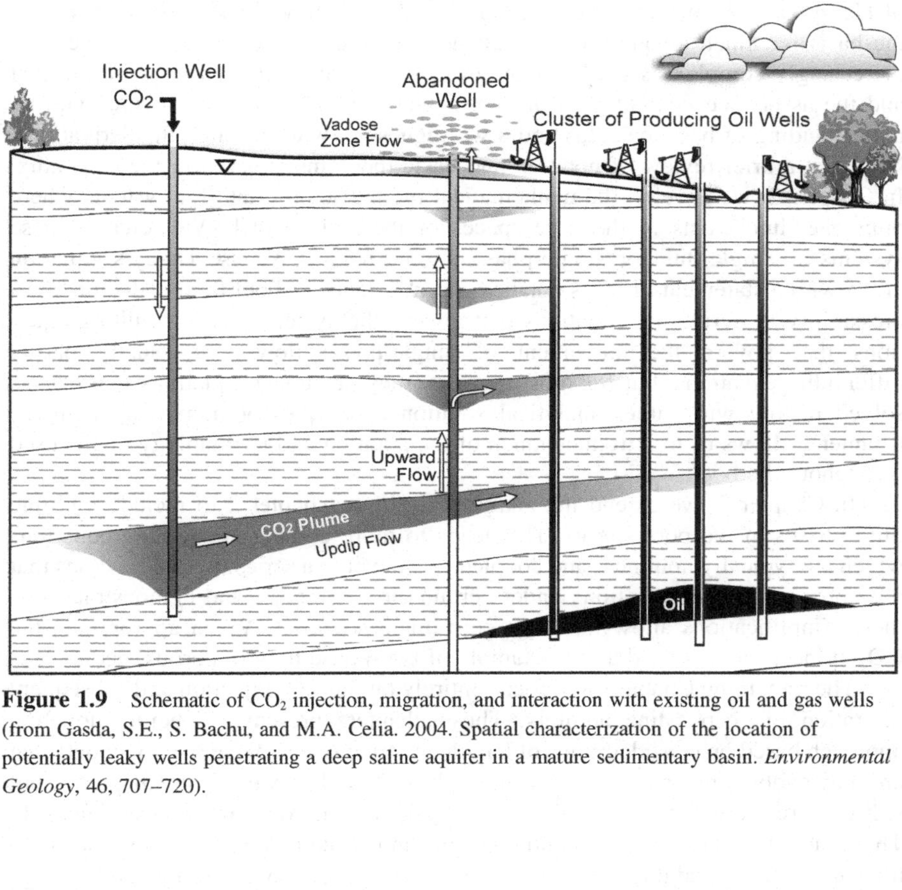

Figure 1.9 Schematic of CO_2 injection, migration, and interaction with existing oil and gas wells (from Gasda, S.E., S. Bachu, and M.A. Celia. 2004. Spatial characterization of the location of potentially leaky wells penetrating a deep saline aquifer in a mature sedimentary basin. *Environmental Geology*, 46, 707–720).

Area of Review that is a natural concept to include in regulatory structures. For all of these aspects of the CO_2 storage problem—movement of the injected CO_2, movement of the displaced brine, possible leakage of one or both of the fluids into and through a complex vertical sequence of formations, and propagation of pressure increases into the formations—we will want to develop appropriate mathematical models and the concomitant methods to solve those mathematics. This book is meant to provide the basic tools to develop, implement, and solve these kinds of models.

1.5 THE APPROACH TAKEN IN THIS BOOK

With the exception of this first chapter, the focus of this book is on mathematical descriptions of fluid movement in the subsurface with a strong focus on CO_2 injection into deep saline aquifers. We have tried to write the book to cover the basics of the processes involved, beginning with a simple description of the experiments of Henry Darcy some 150 years ago, and explaining how those experiments form the basis for much of our current analysis. The mathematical derivations become increasingly complex as the book progresses, but we have tried to write the text and the associated equations so that the material is accessible to anyone with an understanding of basic calculus and a willingness to work through the derivations. Note that a brief review of notation and basic theorems is given in the Appendix. In Chapter 2, we develop the mathematical models appropriate for systems where only one fluid exists in the pore spaces of the soil or rock. We refer to these systems as single-fluid or single-phase porous media. We use Darcy's equation along with a statement of mass balance and information about fluid and solid compressibility to derive the equations that govern the system. Those resulting equations that govern the movement of single-phase flows are usually partial differential equations, but for most systems, they are linear equations that can be solved readily with either analytical solution techniques or numerical methods. Chapter 2 allows us to introduce much of the notation and terminology that is used throughout the book.

In Chapter 3, we extend the analysis begun in Chapter 2 to include multiple fluid phases in the pore spaces. This leads to more complex governing equations which are typically nonlinear and coupled. We develop a series of specific cases that allow for some simplifications under certain circumstances, and demonstrate how these simplifications allow for more convenient solutions. When appropriate, the CO_2–brine system is used as the example of two-phase flow in porous media.

The remaining Chapters 4–6 focus entirely on the CO_2 injection and subsequent migration in deep saline aquifers. Those chapters present a specific modeling approach based on considerations of length and time scales. We develop models that are noticeably simpler than the full multiphase models reviewed in Chapter 3, but still capture the essential behavior of the system at the relevant scales of interest. These models can provide substantial insights into system behavior. They also allow tools to be developed that can simulate large-scale injection, migration, and leakage of both CO_2 and brine, while also accommodating interphase mass exchange and subsequent mass transport of CO_2 as a dissolved component in the aqueous (brine) phase. These models represent all of these processes at the appropriate scales and are sufficiently simple to allow for very efficient solution methods. As such, the methods presented herein can allow realistic CO_2 problems to be analyzed directly, including problems with complex patterns of flow and leakage.

We have included an appendix, meant to provide helpful background and reviews. It includes notation and an overall review of vector and matrix operations as well as a few important results from calculus including Leibnitz' theorem and Gauss' theorem.

1.6 FURTHER READING

At the end of each chapter, we will provide suggestions for further reading on the topic. For Chapter 1 we suggest the following additional readings.

Emissions and Rise of Atmospheric CO_2 Concentrations

CANADELL, J.G., C. LE QUERE, M.R. RAUPACH, C.B. FIELD, E.T. BUITENHUIS, P. CIAIS, T.J. CONWAY, N.P. GILLETT, R.A. HOUGHTON, and G. MARLAND. 2007. Contributions to accelerating atmospheric CO_2 growth from economic activity, carbon intensity, and efficiency of natural sinks. *Proceedings of the National Academy of Sciences of the United States of America*. http://www.pnas.org/cgi/doi/10.1073/pnas.0702737104.

The Future of Coal: Options for a Carbon-Constrained World. The Massachusetts Institute of Technology, 2007.

INTERGOVERNMENTAL PANEL ON CLIMATE CHANGE (IPCC). 2007. *Climate Change 2007: Synthesis Report. Contributions of Working Groups I, II, and III to the Fourth Assessment Report of the IPCC*, Core Writing team, R.K. Pachauri and A. Reisinger eds. Geneve, Switzerland: IPCC.

INTERNATIONAL ENERGY AGENCY (IEA). 2010. *CO_2 Emissions from Fuel Combustion: Highlights, 2010 Edition*. Paris: IEA.

Stabilization Wedges and Technology

PACALA, S. and R. SOCOLOW. 2004. Stabilization wedges: Solving the climate problem for the next 50 years with current technologies. *Science*, 305(5686), 968–972.

SOCOLOW, R.H. and S.W. PACALA. 2006. A plan to keep carbon in check. *Scientific American*, 295(3), 50–57.

Geological Storage of Captured Carbon Dioxide

BACHU, S. and W.D. GUNTER. 2004. Acid-gas injection in the Alberta Basin: A CO_2-storage experience. In: *Geological Storage of Carbon Dioxide*, Vol. 233. S.J. Baines and R.H. Worden eds. London: London Geological Society, Special Publications, pp. 225–234.

INTERGOVERNMENTAL PANEL ON CLIMATE CHANGE (IPCC). 2005. *Special Report on Carbon Dioxide Capture and Storage*. B. Metz et al. ed. Cambridge University Press.

KINTISCH, E. 2007. Making dirty coal plants cleaner. *Science*, 317, 184–185.

TORP, T.A. and J. GALE. 2004. Demonstrating storage of CO_2 in geological reservoirs: The Sleipner and SACS projects. *Energy*, 29, 1361–1369.

WRIGHT, I.W. 2007. The In Salah gas CO2 storage project. Paper IPTC 11326, International Petroleum Technology Conference, Dubai.

The Subsurface Environment

DOMENICO, P.A. and F.W. SCHWARTZ. 1998. *Physical and Chemical Hydrogeology*, Second Edition. New York: John Wiley and Sons.

EPSTEIN, C.M. 1987. Discovery of the aquifers of the New Jersey coastal plain in the nineteenth century. In: *History of Geophysics*, Vol. 3. C.S. Gillmor series ed. Washington, DC: American Geophysical Union, pp. 69–73.

FETTER, C.W. 1994. *Applied Hydrogeology*, Third Edition. New York: Macmillan College Publishing.

FREEZE, R.A. and J.A. CHERRY. 1979. *Groundwater*. Englewood Cliffs, NJ: Prentice-Hall.

KUMP, L.R. 2010. *The Earth System*, Third Edition. San Francisco, CA: Prentice Hall.

PINDER, G.F. and M.A. CELIA. 2006. *Subsurface Hydrology*. Hoboken, NJ: John Wiley and Sons.

Policy and Economic Issues

POLLAK, M.F. and E.J. WILSON. 2009. Regulating geologic sequestration in the United States: Early rules take divergent approaches. *Environmental Science and Technology*, 43, 3035–3041.

RUBIN, E.S., C. CHEN, and A.B. RAO. 2007. Cost and performance of fossil fuel plants with CO_2 capture and storage. *Energy Policy*, 35, 4444–4454.

Chapter 2

Single-Phase Flow in Porous Media

To analyze the movement of CO_2, displaced brine, and perhaps other subsurface fluids, it is important to understand the basic forces that cause fluids to move through porous media, and the equations that allow us to estimate rates of movement. In this chapter we present the basic physics of porous media systems and the associated mathematics that allow for quantitative analysis. The chapter is restricted to systems involving only one fluid phase, with examples including groundwater flow in shallow aquifers and brine flow in deep aquifers. This chapter provides the foundational material on which our analysis of more complex systems, including the CO_2–brine flow system, will be based.

2.1 DARCY'S LAW

One of the most important building blocks for the description of flow in porous media is Darcy's law. In 1856, Henry Darcy published a study of the design of sand filters; part of his work related to a water treatment system for the city of Dijon, France. As part of that work, he performed a number of experiments that allowed him to predict how much water would flow through the sand filters. His experiments form the basis of the equation that was eventually named after him.

Darcy's experiments were simple. He packed a column full of sand, and then made several key empirical observations. They are most easily explained by reference to a generalized schematic drawing of his experiment, as shown in Figure 2.1. In that figure, five important quantities are indicated. The first is the volumetric flow rate of water through the column, denoted by q_{Darcy}. The dimensions associated with q_{Darcy} are volume per time, which we denote by $[L^3T^{-1}]$, where L denotes a length dimension and T denotes time. The second quantity is the height to which the water inside each of the two tubes that penetrate the column rises. These two heights are denoted by h_1 and h_2, with each having dimension of length, or [L]. The third quantity is the cross-sectional area of the column, A $[L^2]$. The fourth quantity is the length

Geological Storage of CO_2: Modeling Approaches for Large-Scale Simulation, First Edition.
Jan M. Nordbotten and Michael A. Celia.

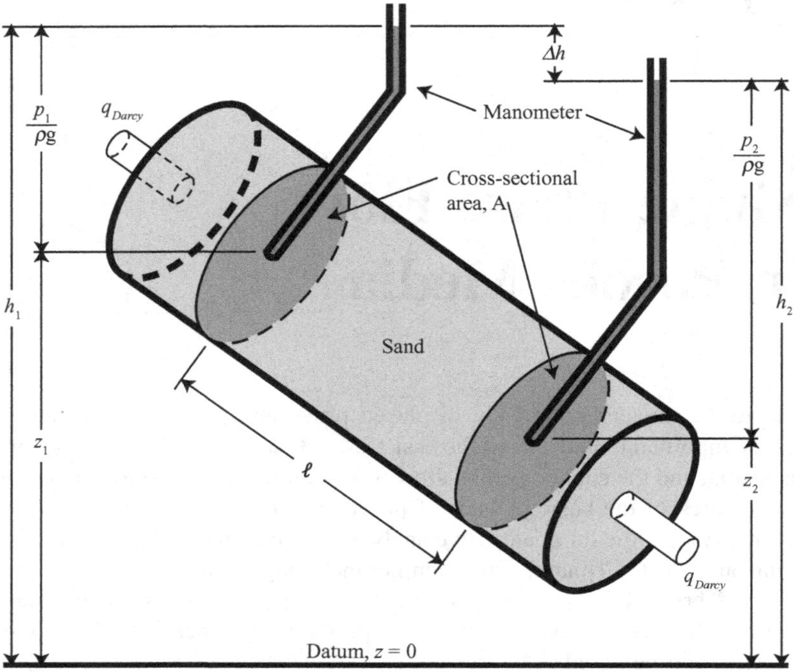

Figure 2.1 Schematic of Darcy's experiment.

of the column, which we denote by the symbol ℓ [L]. And the fifth quantity is the elevation of the point where the end of the small tube penetrates into the column, which is denoted at the two different locations by z_1[L] and z_2[L]. Note that the elevations h and z are measured relative to some specified datum, which is assigned to have the value of zero elevation. Any point above that location has a positive elevation, and any point below it has a negative elevation. Also, note that the subscripts, in this case 1 and 2, denote two different locations along the length of the column.

Darcy performed a number of experiments and made the following general observations: (1) the volumetric flow rate is proportional the difference between h_2 and h_1, (2) the flow rate is proportional to the cross-sectional area A, and (3) the flow rate is inversely proportional to the distance between the measurement points, ℓ. That is,

$$q_{Darcy} \sim \frac{A(h_2 - h_1)}{\ell}, \tag{2.1}$$

where the symbol ~ means "proportional to." If we now define a coefficient of proportionality, κ, then Equation (2.1) can be rewritten as

$$q_{Darcy} = \kappa \frac{A(h_2 - h_1)}{\ell}. \tag{2.2}$$

The coefficient of proportionality was described by Darcy as a "coefficient depending upon the permeability of the sand." We will refer to this coefficient, as used in Equation (2.2), as the *hydraulic conductivity* of the porous medium. Note that this coefficient has dimensions of $[LT^{-1}]$. Also note that Equation (2.2) can be rewritten by dividing both sides of the equation by the area A, and defining a new quantity u, where

$$u \equiv \frac{q_{Darcy}}{A} = \kappa \frac{(h_2 - h_1)}{\ell}. \tag{2.3}$$

The new quantity, u, is a measure of the volumetric flow rate per area of the porous medium, with dimensions $[LT^{-1}]$. Any quantity that is defined on a "per area" and "per time" basis will be referred to as a "flux". Because u is a measure of volume per area per time, we will refer to it as the *volumetric flux* of water through the column. Note that despite the dimensions of length per time, this is not a direct measure of the flow velocity. The velocity is the volume of fluid flowing per area occupied by fluid, while the volumetric flux u is the volume of fluid per total area (which includes both fluid and solid) per time. This is discussed in more detail below and leads to Equation (2.11).

2.1.1 Hydraulic Head

The term h, which appears in the equation and is illustrated in Figure 2.1, is an important quantity in groundwater hydrology and is usually referred to as the *hydraulic head*. In much the same way that heat flows from higher temperature to lower temperature (Fourier's law) and dissolved solute flows from higher concentration regions to lower concentration regions within a fluid (Fick's law), a fluid in a porous medium will flow from regions with higher values of h to regions with lower values of h. This is Darcy's law.

To understand the physical basis of hydraulic head, consider the small tube in Figure 2.1 in which water resides, with one end of the tube connected to the water within the column (say at elevation z_1) and the other end open to the atmosphere. This tube is often referred to as a *manometer*. Because the system is at steady state, water in the manometer is essentially static, meaning it is not moving. A vertical column of water develops a pressure distribution referred to as *hydrostatic*. In a hydrostatic pressure distribution, the density of the fluid, coupled with the gravitational force, leads to an increase in pressure as a function of depth. The pressure within the manometer, which is a force per area $[FL^{-2}]$, is equal to atmospheric pressure at the surface of the water and increases below that surface according to the expression

$$p_{abs}(z) = p_{atm} + \rho g (h_1 - z). \tag{2.4}$$

In this equation, we use the fact that $z = h_1$ corresponds to the water surface in the manometer, and at that point the water pressure is equal to the air pressure, which by definition is atmospheric pressure. In addition, ρ is the fluid density $[ML^{-3}]$, and g is the gravitational acceleration constant [LT^{-2}]. We also denote pressure as the *absolute pressure*, which is denoted by subscript *abs*. Absolute pressure is measured on a scale where a pressure of zero means a total vacuum, with no molecules available to create any pressure—as such it is a true minimum for pressure, an absolute zero.

We may use this equation to determine the pressure at the point where the end of the manometer contacts fluid within the column. At that point, $z = z_1$, and we have $p_{abs}(z_1) = p_{atm} + \rho g(h_1 - z_1)$. By force balance, the pressure of the water at the end of the manometer must equal the pressure of the water in the sand column with which it is in contact. Therefore, the pressure of water in the column at location $z = z_1$ is given by the expression for $p_{abs}(z_1)$. A similar argument gives an analogous expression for the pressure in the column at location $z = z_2$, and this can be generalized to any location within the sand column. If the distance along the sand column is given by the coordinate x, then the general expression for the pressure along the sand column is given by

$$p_{abs}(x) = p_{atm} + \rho g(h(x) - z(x)).$$

A modified pressure scale is often used to redefine where the value of "zero" pressure occurs. This is done by defining a modified pressure, called the *gage pressure*, which is the value of the absolute pressure relative to atmospheric pressure. That is, $p_{gage} = p_{abs} - p_{atm}$. It is customary to drop the subscript gage and simply call the gage pressure "the pressure,", and denote it by the letter p without any subscript. With this notation, we have

$$p = \rho g(h - z).$$

Here we have suppressed that p, h, and z are functions of x. This equation can be rearranged to give an expression for *hydraulic head* h,

$$h = \frac{p}{\rho g} + z. \tag{2.5}$$

Note that this equation says that the hydraulic head is a measure of the pressure at the point of measurement within the column (scaled by ρg) plus the elevation of that point. The scaled value of pressure, $p/\rho g$, is often called the *pressure head*, and has dimension of [L].

We can gain additional insights into the hydraulic head by considering the energy associated with a flowing fluid. The definition of energy per volume from elementary fluid mechanics is given by

$$E \equiv p + \rho g z + \frac{\rho v^2}{2}.$$

One important characteristic of fluid flow in natural porous media is that the flow tends to be very slow relative to flows that are typically seen in open channels. This is due to the large relative importance of friction when fluids flow through small, tortuous flow paths. In a typical groundwater aquifer, a flow rate under natural conditions of 10 cm/day is reasonably "fast." This means that in the definition of energy, the velocity is very small, and the velocity-squared term is almost always negligibly small. Therefore, for subsurface flow systems, the kinetic energy term can almost always be neglected, and the energy content of the fluid becomes $E \approx p + \rho g z$. We observe that this measure of fluid energy is simply a scaled version of the hydraulic head of Darcy, that is, $h = E/\rho g$. E has dimension of energy per volume, while h has dimension of energy per force which is the same as length.

2.1.2 Hydraulic Conductivity and Permeability

The hydraulic conductivity is one of the most important properties of porous media because it indicates the ease with which fluids can flow through the material. It is a function of both the porous medium and the fluid flowing through it. A Darcy-type column experiment that uses molasses or some other highly viscous fluid will produce slower flow rates, and therefore a lower value of hydraulic conductivity, than identical experiments using water or a less viscous fluid. The particular dependence of hydraulic conductivity on the fluid properties may be derived by dimensional analysis and the result can be written as follows:

$$K = \frac{k\rho g}{\mu}. \tag{2.6}$$

In Equation (2.6), μ is the dynamic viscosity of the fluid [$ML^{-1}T^{-1}$] and k is a coefficient that depends on the porous medium but not on the fluid [L^2]. This coefficient k is called the *intrinsic permeability*, or often just the *permeability*. It is a very important property of the rock or soil. Note that it has dimension [L^2], so typical units are m^2 or cm^2, or derived units called *Darcy* and *milliDarcy*, where one Darcy is equal to roughly 10^{-12} m^2. Typical ranges for values of both hydraulic conductivity and permeability are given in Table 2.1, using different units.

One of the interesting aspects of flow in porous media is that the porous medium may allow fluid to flow more easily in one direction than another. This happens, for example, when the material exhibits small-scale layering such that within a given sample of the material, there are thin layers of different kinds of materials. In these cases, it is usually noticeably more difficulty to flow perpendicular to the layering than parallel to it. If the layering is horizontal, then the more difficult flow direction will be the vertical direction (see Box 2.1). When a parameter or property changes value depending on the direction being considered, the system is referred to as *anisotropic*. Conversely, when there are no directional differences, the material or system is called *isotropic*. When a property or parameter such as permeability changes as a function of spatial location, the system is referred to as *heterogeneous*. Conversely, when a system is spatially uniform, it is *homogeneous*. For geological settings, there is a general trend that permeability decreases with depth, due to higher levels of compaction. This trend is superimposed on the natural layering of the medium which comes from the sedimentary processes that created the geological structures. Note that a homogeneous system can be either isotropic or anisotropic;

BOX 2.1 *Layered Media and Anisotropic Effective Hydraulic Conductivity*

A layered material will have an anisotropic hydraulic conductivity when the scale of averaging is large relative to the layering. Consider the simple layered system shown in Figure 2.A. We will consider flow both parallel to the layering and perpendicular to the layering.

For flow parallel to the layering, we seek to determine the effective hydraulic conductivity that captures the bulk effect of the layering. That is, if flow is in the direction of the layering, say from left to right, then we seek a single conductivity value that properly represents the total flow through all the layers. If the hydraulic head along the left side is denoted by h_L and the head along the right boundary is h_R, then we seek an effective hydraulic conductivity value $\kappa_{\|}^{eff}$ such that the integrated flow $u_{\|}$ within the layered system is represented by the equation

$$u_{\|} = \kappa_{\|}^{eff} \frac{h_L - h_R}{d}.$$

To relate $\kappa_{\|}^{eff}$ to the conductivity values of each of the individual layers, we observe that the total volumetric flow is given by the following expression,

$$q_{Darcy} = u_{\|} A = \kappa_{\|}^{eff} \frac{h_L - h_R}{d}\left(\sum_i b_i\right) d = \sum_i \kappa_i \frac{h_L - h_R}{d} b_i d.$$

This immediately shows that the effective conductivity is the weighted arithmetic average of the layer conductivity values, with the weights related to the layer thicknesses,

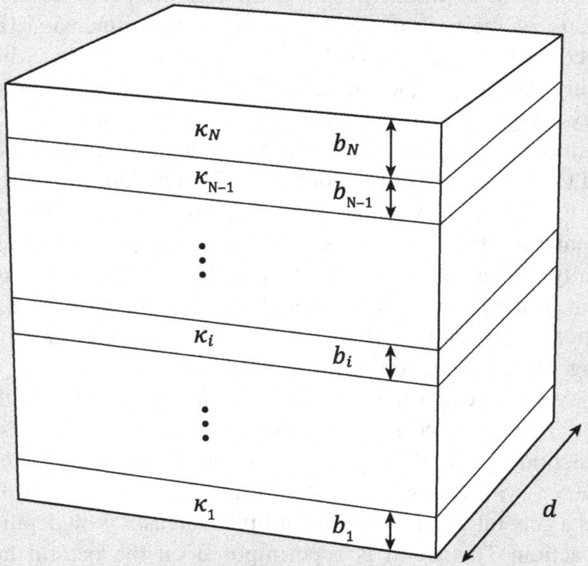

Figure 2.A Sketch and notation of layered system.

$$\kappa_{\parallel}^{eff} = \frac{\sum_i \kappa_i b_i}{\sum_i b_i}.$$

Flow perpendicular to the layers, $u_{\perp}$, is a bit more involved. Now assume the flow is from top to bottom in the figure, and let the hydraulic head along the top and bottom boundaries be given by h_T and h_B, respectively. We observe that conservation of mass requires that the flow across each layer must be the same, which implies the following constraint on the effective perpendicular conductivity $\kappa_{\perp}^{eff}$,

$$\kappa_i \frac{\Delta h_i}{b_i} = u_i = u_{\perp} = \kappa_{\perp}^{eff} \frac{h_T - h_B}{\sum_i b_i}$$

where Δh_i represents the head drop across layer i. This equation is constrained by the fact that the total drop in head across all layers is equal to the head drop from top to bottom, that is,

$$\sum_i \Delta h_i = h_T - h_B.$$

When combined, these last two equations yield the following expression for the effective hydraulic conductivity,

$$\kappa_{\perp}^{eff} = \frac{\sum_i b_i}{\sum_i \frac{b_i}{\kappa_i}}$$

which is a weighted harmonic average of the individual hydraulic conductivities.

Flows parallel to the layering lead to a weighted arithmetic mean while flows perpendicular to the layering lead to a weighted harmonic mean. If all layers have the same conductivity values (i.e., the material is homogeneous), then the two values are identical. Whenever the values are not all identical, then the arithmetic average will always be larger than the harmonic average. As such, flow parallel to layering is always "easier" than flow perpendicular to the layering, and materials with layered structures on scales smaller than the averaging (REV) length scale will always be anisotropic.

similarly, a heterogeneous system can be either isotropic or anisotropic. Furthermore, the degree to which a system is heterogeneous or anisotropic often depends on the spatial scale used to define the properties of interest.

While we will not dwell on this particular concept here, it is worth remembering this fact when we address the issue of CO_2 injection and migration, because there we will need to address length scales and their impact on the equations we write.

2.1.3 Extensions of Darcy's Law

While Darcy derived the algebraic relationships given in Equations (2.1) to (2.3), in the subsequent 150 years, there have been many extensions to what we now call

Table 2.1 Permeability and Conductivity Values for Various Soils (Taken from Freeze and Cherry, 1979)

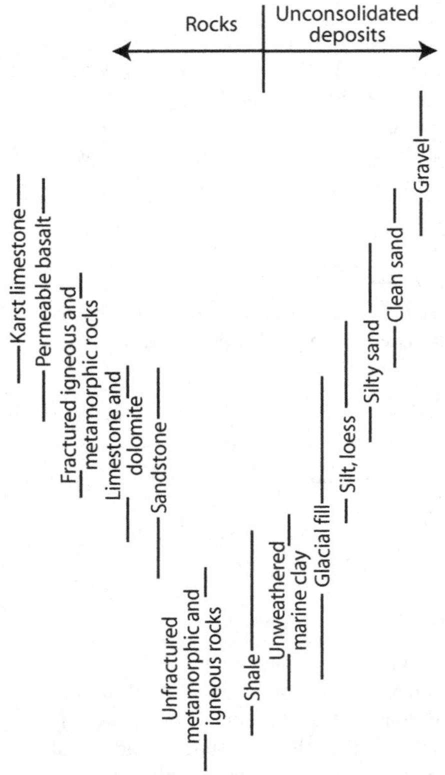

k (darcy)	k (m²)	κ (cm/s)	κ (m/s)
10^5	10^{-7}	10^2	1
10^4	10^{-8}	10	10^{-1}
10^3	10^{-9}	1	10^{-2}
10^2	10^{-10}	10^{-1}	10^{-3}
10	10^{-11}	10^{-2}	10^{-4}
1	10^{-12}	10^{-3}	10^{-5}
10^{-1}	10^{-13}	10^{-4}	10^{-6}
10^{-2}	10^{-14}	10^{-5}	10^{-7}
10^{-3}	10^{-15}	10^{-6}	10^{-8}
10^{-4}	10^{-16}	10^{-7}	10^{-9}
10^{-5}	10^{-17}	10^{-8}	10^{-10}
10^{-6}	10^{-18}	10^{-9}	10^{-11}
10^{-7}	10^{-19}	10^{-10}	10^{-12}
10^{-8}	10^{-20}	10^{-11}	10^{-13}

"Darcy's law." In this section we will consider a few of these, and then add a few more in the next chapter.

The first extension is a replacement of the algebraic differences in Equation (2.3) with a differential expression derived by allowing the measurement points associated with h_1 and h_2 to become arbitrarily close to one another. If we assume the column is aligned with the vertical (z) direction, and if we assume hydraulic head $h(z)$ is a well-behaved function, then we can take the limit as the distance goes to zero to find the following differential form of Darcy's equation:

$$u = -\kappa \frac{dh}{dz}. \tag{2.7}$$

Note that the minus sign is consistent with the idea that fluids flow in the direction from higher hydraulic head to lower hydraulic head.

So far, we have only considered the Darcy experiment, which involves a column and is therefore restricted to flows along the axial direction of the column. In general, the volumetric flow is a vector quantity, which in three-dimensional space will have three different components making up the overall vector $\boldsymbol{u}$. We will write the vector using its components, u_i, and the appropriate unit vectors, $\boldsymbol{e}_i$, such that $\boldsymbol{u} = [u_1; u_2; u_3] = u_1\boldsymbol{e}_1 + u_2\boldsymbol{e}_2 + u_3\boldsymbol{e}_3$. Note that we use boldface type to denote a vector.

Let us now consider Darcy's law for a three-dimensional flow field. If the movement is governed by the hydraulic head, and flow occurs from higher values of h to lower values of h, the appropriate multidimensional version of the spatial derivative is the gradient, which in Cartesian coordinates we write as

$$\nabla h = \frac{\partial h}{\partial x_1}\boldsymbol{e}_1 + \frac{\partial h}{\partial x_2}\boldsymbol{e}_2 + \frac{\partial h}{\partial x_3}\boldsymbol{e}_3.$$

We then extend the one-dimensional version of Darcy's law to three dimensions, for isotropic hydraulic conductivity fields, by writing the flow vector as follows:

$$\boldsymbol{u} = -\kappa \nabla h.$$

When the hydraulic conductivity is anisotropic, we need to have a procedure to include values of the hydraulic conductivity that are dependent on direction. When we wish to relate one vector ($\boldsymbol{u}$) to another (∇h), the field of linear algebra tells us to use a matrix operator, which in three-dimensional space is a 3×3 matrix. That is, when the hydraulic conductivity is anisotropic, we need to form a conductivity *matrix* that will multiply the gradient of hydraulic head to give the flow vector. If we again use boldface to indicate a matrix, the result is the following equation:

$$\boldsymbol{u} = -\boldsymbol{\kappa} \nabla h. \tag{2.8}$$

We will avoid most discussion of the hydraulic conductivity matrix (also referred to as the hydraulic conductivity "tensor"), except to note that at any given point in space, the anisotropic material will have a specific direction along which the hydraulic conductivity is a maximum and a direction along which it is a minimum. These directions, referred to as the *principal directions*, are orthogonal to one another. If the coordinate system $\boldsymbol{x} = (x_1, x_2, x_3)$ is aligned with these "principal directions," then the hydraulic conductivity matrix will be diagonal, with $\kappa_{ij} = 0$ for $i \neq j$. In that case, the equations for flow simplify so that $u_i = -\kappa_{ii}(\partial h / \partial x_i)$ for $i = 1, 2, 3$. In most of what follows, if anisotropy is included in the analysis, we will assume that the coordinate axes align with the principal axes of the hydraulic conductivity tensor. Note that given the relationship between hydraulic conductivity and intrinsic permeability (Equation (2.6)), and given that fluid density, fluid viscosity, and the gravitational constant are all scalars, the directional nature of the hydraulic conductivity must derive from the directional nature of the intrinsic permeability. Therefore, we speak of anisotropic (intrinsic) permeability analogously to anisotropic hydraulic conductivity, and observe that anisotropy arises from the rock properties and not from the fluid properties. Note that we will use vector notation as outlined above consistently throughout the book; a concise review of this notation is provided in the Appendix.

When deriving Darcy's law from first principles, we get the final extension of Darcy's equation for single-phase flow:

$$\boldsymbol{u} = -\frac{\boldsymbol{k}}{\mu}(\nabla p + \rho g \nabla z). \tag{2.9}$$

This equation is related to our previous versions of Darcy's law through a combination of Equations (2.5), (2.6), and (2.8). Assume Equation (2.6) is written in the more general matrix notation. Then we can substitute the expanded representation for both $\boldsymbol{\kappa}$ and ∇h into Darcy's equation. We then see that the earlier versions of Darcy's law only correspond to the more general form given in Equation (2.9) in the case where density is constant. When considering Equation (2.9), it is helpful to observe first that the gradient of the vertical coordinate z (which may or may not coincide with the general coordinate x_3) may be denoted by the unit vector along the vertical direction, $\boldsymbol{e}_z$ (note that any unit vector may be written equivalently as the gradient of its coordinate, for example $\nabla x_1 = \boldsymbol{e}_1$), and second, that the gravitational acceleration vector may be defined as $\boldsymbol{g} = -g\boldsymbol{e}_z$, where the vertical coordinate is taken with positive direction upward. With this final definition, we may write Darcy's law in the following equivalent forms, which we will use often in the subsequent chapters:

$$\boldsymbol{u} = -\frac{\boldsymbol{k}}{\mu}(\nabla p + \rho g \nabla z) = -\frac{\boldsymbol{k}}{\mu}(\nabla p + \rho g \boldsymbol{e}_z) = -\frac{\boldsymbol{k}}{\mu}(\nabla p - \rho \boldsymbol{g}). \tag{2.10}$$

Note that throughout the book, we will follow the convention used here that the vertical direction is always denoted by the coordinate z, and it is always taken to be positive upward.

In this section we have used the original Darcy experiments as a starting point, and then generalized the resulting relationship for flow in porous media. This is generally consistent with the historical development of these extensions, and in many ways is the easiest approach to explain the underlying meaning of the terms and how they relate to one another. However, there are many studies that have attempted to derive Darcy's Law from first principles, where the equations of fluid dynamics are written to describe fluid flow in the pore spaces of the porous medium, and then mathematical averaging operators are defined and applied to arrive at an equation that looks like Darcy's law. When these general principles are applied, the usual result is an equation that looks very much like Equation (2.10). In that regard, we may view Equation (2.10) as a more general statement of the relationship between fluid flow and the driving forces for that flow. With this approach, the simpler versions of Darcy's law, like Equation (2.8), only result from special cases. In particular, we note that Equation (2.8) can only be derived from Equation (2.10) when the fluid density is either constant (which we assumed in our earlier derivation of Darcy's law) or is a function of only fluid pressure. If the fluid density depends on fluid composition (concentration of dissolved components), temperature, or any other factors, then a unique fluid potential (E or h) cannot be derived. The simplified version of Darcy's law, Equation (2.8), using the fluid potential, is often used in the broad field of groundwater hydrology, while the more general form given in Equation

(2.10) is usually used in petroleum reservoir engineering. For most of the remainder of this chapter, we will use the simpler version with the hydraulic head as a primary unknown. In later chapters, when we deal specifically with CO_2 injection and fluid migration in deep geological formations, we will use the pressure-based form in Equation (2.10) and avoid explicit use of the fluid potential in our formulations.

2.2 CONSERVATION LAWS AND GOVERNING EQUATIONS

While Darcy's law explains how a fluid moves in a porous medium, it is not sufficient to analyze general flow problems because it does not constrain the system sufficiently to allow for a unique solution. A second important equation to describe fluid movement in porous media is a mathematical statement of the principle of conservation of mass. The underlying idea is to identify a specific region of space and to make the simple observation that over some time period, the change of mass of a particular substance within the identified region of space is equal to the amount of mass that enters (or leaves) through the entire boundary of the region, plus (or minus) any mass that appears (or disappears) from within the region without passing through its boundaries. This simple statement forms the basis for the equation of mass conservation.

2.2.1 Mass Conservation

In order to apply the concept of mass conservation, and to derive a set of equations that can be solved for variables of interest, we begin this section with definitions of the properties and functions we will need to describe flow in porous media. These are all motivated by the fact that the material we are dealing with is geometrically highly complex and cannot be characterized explicitly (or "deterministically") because of its geometric complexity. Figure 2.2 shows a visualization of a digitized porous medium (a sandstone) where complex, tortuous flow pathways exist within the pore space of the material. These flow paths tend to have characteristic width on the order of a fraction of one micron to tens or perhaps hundreds of microns (1 micron = 1 micrometer = 10^{-6} meters). Because we cannot reasonably resolve this scale, either observationally or computationally, we instead define averages over length scales that are more convenient. The length scale that we choose here is what we refer to as a *representative elementary volume* (REV) associated with the laboratory scale, or REV_{lab}. This is a scale which is large enough to allow for meaningful (statistically representative) averages to be defined over the pore space, and on which practical laboratory measurements can be made. An example of this length scale is the size of a core sample of soil taken in the field and brought to the laboratory for analysis. The characteristic length scale of the REV_{lab} is on the order of one centimeter to a few tens of centimeters. This characteristic length is denoted by ℓ_{lab}.

The simplest characterization of the pore space is a geometric measure of the fraction of the overall sample volume that is occupied by pore space. This fraction

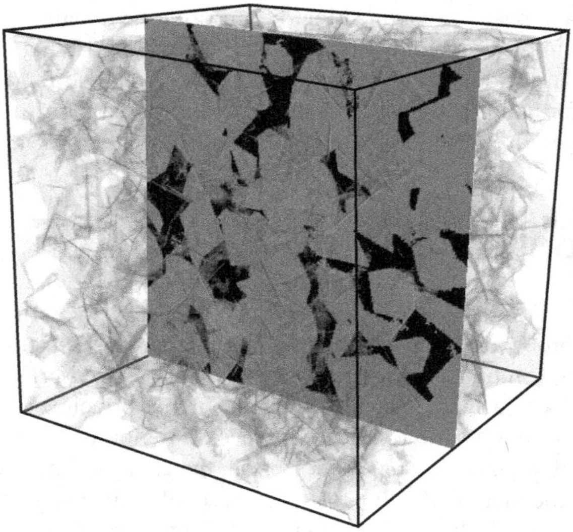

Figure 2.2 Illustration of complex geometry of a geological porous material (Courtesy of Numerical Rocks A/S).

is referred to as the *porosity*, and will be denoted by ϕ. We note that the porosity may be a function of spatial location (i.e., the porous medium may be heterogeneous), it may be a function of time (although such temporal changes usually tend to be very small), and it is dependent on the overall averaging length scale. That is, the porosity function may be written as $\phi(\boldsymbol{x}, t; \ell_{lab})$, showing explicit dependence on space and time, parameterized by the averaging scale ℓ_{lab}. If we imagine the averaging volume to be represented by a specific shape (a sphere or a box) with characteristic length ℓ_{lab}, then the porosity of the material in that volume will be defined as the single (averaged) value, and assigned to the centroid of the volume. That is, the value of porosity assigned to point $\boldsymbol{x}$ is the porosity in the averaging volume with characteristic length ℓ_{lab} whose centroid resides at $\boldsymbol{x}$. The resulting averaged functions are thereby well defined and are assumed to be smooth in space, so they can be differentiated or integrated in the usual ways. Note that with this spatial averaging, the explicit geometry of pore and solid disappears and at any point in space, there is a well-defined value of porosity independent of whether the actual physical location of $\boldsymbol{x}$ corresponds to pore space or solid. This is analogous to defining fluid properties like density and viscosity in fluid mechanics whereby discrete molecular interactions are eliminated from explicit consideration. Since parameters defined in this way are now spatially continuous, the laboratory scale is sometimes referred to as the "continuum scale" for porous media flows.

In addition to porosity, we can define other averaged variables. Fluid density may be averaged over the REV_{lab}, and the flow vector $\boldsymbol{u}$ can be averaged so that it corresponds to the values measured, for example, in the laboratory-scale Darcy-type experiments. Recall that the volumetric flux measured in the Darcy experiments is defined as volume of fluid per total area per time, where "total area" means that the full cross-sectional area is used, including both the pores and the solid. This is consistent with our use of the term "flux." The actual averaged fluid velocity is the volume of fluid flowing through a particular cross section per area occupied by that fluid (instead of the total area). This means that the fluid velocity vector, which we will denote by $\mathscr{v}$, is a scaled version of the flux vector $\boldsymbol{u}$ where the scale factor is the fraction of the total space occupied by the fluid, which in this case is the porosity. That is,

$$\mathscr{v} = \frac{\boldsymbol{u}}{\phi}. \tag{2.11}$$

Given this set of functions corresponding to volume averages over the length scale ℓ_{lab}, which can be measured readily in the laboratory, we can now derive the equation for mass conservation in terms of these functions. For this we consider an arbitrary volume in space (note that this volume is *not* related to the averaging volume which has a specified characteristic length ℓ_{lab}), and we use the definition of mass conservation stated in the first paragraph of this section. That is, any change of mass in the volume is balanced by the net mass flow into the volume through its boundaries and by any mass added to the volume not associated with the boundary fluxes. If we denote the volume by Ω and the boundary of the volume by $\partial\Omega$, then the mathematical statement of conservation may be written in integral form as follows,

$$\int_{\Omega} \frac{\partial \mathscr{m}}{\partial t} dV = -\oint_{\partial\Omega} \mathscr{f} \cdot \mathbf{v}_n dA + \int_{\Omega} \mathscr{r} dV. \tag{2.12}$$

In Equation (2.12), $\mathscr{m}$ represents a measure of mass per total volume of porous medium, $\mathscr{f}$ is the mass flux vector, $\mathbf{v}_n$ denotes the unit vector in the direction normal to the surface $\partial\Omega$, directed outward (hence the negative sign in front of the surface integral), and $\mathscr{r}$ represents any sources or sinks of mass within the volume, in units of mass per volume per time. If the sources and sinks are zero, or correspond to known external sources, then $\mathscr{m}$ is a locally conserved quantity for the system and we refer to Equation (2.12) as a *conservation law*. Otherwise, $\mathscr{r}$ will represent changes internal to the system, in which case we will refer to Equation (2.12) as a *balance equation* or *transport equation*. In the case of mass conservation,

$$\mathscr{m} = \rho\phi, \; \mathscr{f} = \rho\boldsymbol{u}, \text{ and } \mathscr{r} = \psi, \tag{2.13}$$

where ψ represents an external source or sink term of mass. The surface integral term in Equation (2.12) may be replaced by a volume integral using Gauss's theorem (also called the divergence theorem—see the Appendix). Therefore, Equations (2.12) and (2.13) may be combined with the divergence theorem to write

$$\int_{\Omega} \left(\frac{\partial \mathscr{m}}{\partial t} + \nabla \cdot \mathscr{f} - \mathscr{r} \right) dV = \int_{\Omega} \left(\frac{\partial \rho\phi}{\partial t} + \nabla \cdot (\rho\boldsymbol{u}) - \psi \right) dV = 0. \tag{2.14}$$

Because the integral in Equation (2.14) must hold for any arbitrary closed volume Ω, it must be the case (given sufficient smoothness of the terms involved) that the integrand itself is zero. Therefore, we obtain the differential equation

$$\frac{\partial(\rho\phi)}{\partial t} + \nabla \cdot (\rho\boldsymbol{u}) = \psi. \tag{2.15}$$

We will take Equation (2.15) to represent the general mass conservation equation for flow of a single fluid in a porous medium. Along with Darcy's law (Equation (2.8) or Equation (2.10)), it forms the basis of much of the mathematical analysis related to flow in porous media (see Box. 2.2).

We are sometimes interested in not only the overall fluid phase but in the movement in one or more of the components that make up that phase. For example,

BOX 2.2 *Volume-Based Capacity Estimates*

The simplest estimates of potential storage are based on the available volume in geological formations. We can therefore start estimating storage capacity already with the fundamental concepts of volume balance. We will quickly go through a few simple calculations in order to better understand the magnitudes involved. In particular, all volume-based capacity estimates are based on integration over the pore volume

$$M_\Sigma = \int_\Omega \rho\phi dV \approx \rho \int_\Omega \phi dV.$$

A relatively large aquifer may have a spatial extent on the order of 2500 km^2, with an average thickness of 40 m and porosity of 10%. This gives a fluid volume of 10,000 km^3, which we will use as our characteristic volume for aquifer storage, and denote by V_{Aqf}.

For our characteristic volume, we can derive several basic estimates, depending on how efficiently we can utilize the pore space. The most optimistic estimate assumes that the whole pore space is available for CO_2 storage. This entails displacing (and handling) all of the resident fluids, which are likely to be salty brines. If CO_2 is compressed to a density of 700 kg/m^3, this gives a storage capacity of $M_\Sigma \approx$ 7000 Gt CO_2 per V_{Aqf}. Recall that the global yearly emissions of CO_2 are on the order of 30 Gt CO_2. We will see in following chapters that achieving anywhere close to this storage capacity is almost impossible due to residual water and accessibility issues, and furthermore may be undesirable from a leakage risk perspective.

A less optimistic estimate considers only CO_2 as a dissolved component in the resident brine. Using a mass fraction of CO_2 in brine of 4%, this implies a storage capacity of about $M_\Sigma \approx$ 400 Gt CO_2 per V_{Aqf}. This estimate is optimistic in that it does not address the practical and economical issues of accessing all the pore space. On the other hand, it is pessimistic in not considering any CO_2 in a separate phase. On balance, this second estimate comes closer to an order of magnitude estimate, in the absence of economic barriers, than the first alternative provided.

For comparison, the current injection in the Utsira formation has an observed footprint based on seismic imaging of about 4 km^2 after 14 Mt CO_2 injection. Here, the thickness of the aquifer is on the order of 200 m with a porosity on the order of 30%. This implies a current local storage efficiency of almost 600 Gt CO_2/V_{Aqf}.

Both estimates provided above can be augmented by factors that take into account interaction between fluid phases, local geometry, heterogeneity of the formation, along with several engineering and economical thresholds. These factors, and therefore also storage capacity estimates, will inherently be site-specific. Addressing these site-specific issues can be achieved through the use of large-scale multiphase flow modeling, as we will see in the later chapters of this book.

in groundwater problems, we may be interested in the transport of a dissolved contaminant such as an organic solvent. We can use the formalism of Equation (2.12) to write the conservation equation for a component within the fluid phase. If we denote a component within the fluid phase by index *i*, then the *concentration* of the component may be defined on a mass basis as the *mass fraction*, which we denote by m_f^i and define as the ratio of the mass of component *i* to the total mass of the fluid phase, which is denoted by the subscript *f*. A dissolved component may be transported by the bulk flow of the fluid phase, which we refer to as *advective transport*, or it may be transported by non-advective mechanisms like molecular diffusion, where concentration gradients can drive a mass flux following Fick's law. If all non-advective fluxes are identified by the vector $\boldsymbol{j}^i$, then the mass term $\mathscr{m}$ and the flux vector $\mathscr{f}$ may be defined for component transport as follows,

$$\mathscr{m} = \rho\phi m_f^i, \ \mathscr{f} = \rho \boldsymbol{u} m_f^i + \boldsymbol{j}^i \text{ and } \mathscr{r} = \psi^i.$$

This gives the basic component conservation equation for single-fluid flow in a porous medium,

$$\frac{\partial\left(\rho\phi m_f^i\right)}{\partial t} + \nabla\cdot\left(\rho \boldsymbol{u} m_f^i + \boldsymbol{j}^i\right) = \psi^i, \tag{2.16}$$

where the source/sink term ψ^i represents the rate at which mass of component *i* is added to (or removed from) the fluid phase. Note that when we include chemical or biological reactions, the component mass in the fluid phase is no longer conserved according to our earlier definition of "conservation" equations (as opposed to "balance" equations) because the source or sink terms now include internal reactions as well as external sources or sinks. In this case, Equation (2.16) is simply a mass balance equation, where source/sink term ψ^i represents the mass rate per volume of porous medium at which chemical or biological reactions add (or remove) component *i* from the fluid phase. This exemplifies the distinction between a conservation equation and a balance equation: The conservation equation has a known source term ψ which is external to the system, while in the balance equation, additional information, and functional dependencies, about the source term needs to be provided.

We note that for the CO_2 problem, because injected CO_2 is slightly miscible with brine, some of the CO_2 will dissolve into the brine and be transported, as a dissolved component, with the brine. This leads to a component transport problem that can be important on long time scales. A more thorough discussion is included in Chapters 3 and 4.

We also note that the non-advective flux term includes not only molecular diffusion but also mechanical mixing caused by velocity variability on length scales below the scale used to define the average velocity. If this is the laboratory scale, then the subscale variations occur on scales not larger than centimeters. These inhomogeneities of velocity cause additional mixing and spreading of components, with this process usually called *mechanical dispersion*. When field-scale problems are analyzed, and velocities are approximated on much larger averaging volumes, then subscale velocity variations can become relatively large. This large-scale variability of velocity, which is not captured through the average velocity in the advection term, is often referred to as *macro-dispersion*. It should be clear that the magnitude of the dispersion term depends on the scale of resolution of the average velocity. As such, it is not surprising that the magnitude of dispersion depends on the scale of resolution.

2.2.2 Governing Equations with Compressibility

Examination of Equation (2.15) indicates that there are at least three unknowns in the equation: u_1, u_2, u_3. In addition, if either the density or the porosity changes in time, then these also need to be determined. With more than one unknown, the single mass conservation equation cannot be solved without being combined with other equations. To derive a set of equations that can be solved, we use Darcy's law and an expression for compressibility.

Darcy's law is already written in the form using hydraulic head (Equation (2.8)) or pressure (Equation (2.10)). Either of those equations can be used to substitute for the flow vector $\boldsymbol{u}$. In that way, the vector of unknowns $\boldsymbol{u}$ is replaced by a single scalar unknown, either the hydraulic head or the pressure. In the case of steady-state flow, time-derivatives are zero. In that case, we have one equation (Darcy's law inserted into mass conservation) for one unknown, and the system can be solved. However, in general, we also need to consider temporal changes.

To deal with the time derivative, we need to consider possible compressibility of both the fluid and the solid matrix. The first leads to possible changes in fluid density, and the second leads to possible changes in porosity. While a number of factors can result in changes in fluid density, including dependence on composition or temperature, at this point we will only consider changes in density due to changes in fluid pressure. In that regard, we define a measure of fluid compressibility as follows,

$$c_f \equiv \frac{1}{\rho}\frac{d\rho}{dp}. \tag{2.17}$$

Equation (2.17) states that if the fluid pressure is changed by some amount, the density will change according to the compressibility coefficient c_f. Typical values of compressibility for water are around 5×10^{-10} Pa^{-1} which implies that water is only slightly compressible.

When considering a system of soil and fluid—assume the fluid is groundwater—a point at some depth z will have a total weight above it based on the mass of both the solid and the fluid. That weight is supported by a combination of fluid pressure and what is referred to as *effective stress*, which is the stress carried by the solid through grain-to-grain contacts and which we denote by σ_{eff} (see Box 2.3). If the water pressure decreases, then more of the weight needs to be carried by the effective stress (since for confined formations, the stress induced by the

BOX 2.3 *Effective Stress*

The concept of effective stress is usually defined from a relationship of the form

$$\sigma_{eff} = \sigma - a_p p$$

where σ is the total exterior stress of the porous medium and fluid. Since stress in general is a tensor, this expression is a tensor relationship, and the fluid pressure is then interpreted as an isotropic tensor. The coefficient of proportionality, a_p, designates the proportion of the total stress carried by the fluid. Depending on the physical properties of the porous medium, the coefficient a_p may depend on either porosity or compressibility, or both. In this book, we will not emphasize these nuances, and simply consider the coefficient $a_p = 1$, and further neglect the tensorial nature of stress, and simply represent effective stress by a scalar function. Additional information on effective stress can be found in the references.

geology above the formation is constant, and therefore $p + \sigma_{eff} = \sigma_\Sigma = constant$, where σ_Σ denotes the vertical component of total stress, and constant total stress implies that $d\sigma_{eff} = -dp$). This increase in effective stress can lead to changes in porosity through consolidation of the solid matrix. This consolidation is usually related to the change in effective stress (and therefore the fluid pressure) through a compressibility coefficient for the solid, c_ϕ, defined as

$$c_\phi \equiv -\frac{d\phi}{d\sigma_{eff}} = \frac{d\phi}{dp}.$$

To see how these compressibility coefficients enter into the mass balance equation, we expand the time derivative in Equation (2.15) using the chain rule of differentiation, under the assumption that both density and porosity are functions of fluid pressure.

$$\begin{aligned}\frac{\partial(\rho\phi)}{\partial t} &= \rho\frac{\partial\phi}{\partial t} + \phi\frac{\partial\rho}{\partial t} = \rho\frac{d\phi}{dp}\frac{\partial p}{\partial t} + \phi\frac{d\rho}{dp}\frac{\partial p}{\partial t} = \rho c_\phi\frac{\partial p}{\partial t} + \phi\rho c_f\frac{\partial p}{\partial t} \\ &= \rho\left(c_\phi + \phi c_f\right)\frac{\partial p}{\partial t} = \rho c_\Sigma\frac{\partial p}{\partial t}.\end{aligned} \tag{2.18}$$

In Equation (2.18) we have defined the total compressibility coefficient, c_Σ, given by

$$c_\Sigma \equiv c_\phi + \phi c_f.$$

If we use the substitution of Equation (2.18) for the time derivative, and then use Equation (2.10) to substitute for the flow vector, we can rewrite the mass conservation equation in differential form (Equation (2.15)) as

$$\rho c_\Sigma\frac{\partial p}{\partial t} - \nabla\cdot\left(\rho\frac{k}{\mu}(\nabla p - \rho g)\right) = \psi. \tag{2.19}$$

Notice that Equation (2.19) is now one equation with only one unknown, that being the pressure $p(\boldsymbol{x}, t)$. We often assume that the spatial changes in density can be neglected, which means that the term $\boldsymbol{u} \cdot \nabla \rho$ is negligible. Under this assumption, Equation (2.19) can be rewritten as

$$c_{\Sigma}\frac{\partial p}{\partial t} - \nabla \cdot \left(\frac{\boldsymbol{k}}{\mu}(\nabla p - \rho \boldsymbol{g})\right) = \frac{\psi}{\rho}. \tag{2.20}$$

Equations (2.19) and (2.20) are single-phase three-dimensional flow equations in terms of pressure. As they are derived from the mass conservation equation but stated for a nonconserved variable (pressure), we will refer to them as three-dimensional mass balance equations.

A similar equation can be written with hydraulic head as the primary unknown. To do this, we observe that $\partial p/\partial t \approx \rho g(\partial h/\partial t)$, and we use the hydraulic head-based version of Darcy's law, Equation (2.8), to rewrite Equation (2.20) as follows,

$$S_s\frac{\partial h}{\partial t} - \nabla \cdot (\boldsymbol{\kappa} \nabla h) = \upsilon. \tag{2.21}$$

Here, the *specific storativity*, $S_s \equiv c_{\Sigma}\rho g$, has been introduced. Note that Equation (2.21) has units of volume per volume per time, and we have defined the Greek letter upsilon (υ) to represent volumetric source terms:

$$\upsilon \equiv \frac{\psi}{\rho}.$$

Equations (2.19), (2.20) and (2.21) are three-dimensional equations for fluid flow for a single-fluid (single-phase) porous medium. To actually solve these equations, the specific spatial and temporal domains within which these equations apply must be specified. For these second-order-in-space equations, we usually need to specify the location of the spatial boundary and one boundary condition at every point along the boundary. We also need to specify one initial condition at every point within the domain. These are standard requirements for the solution of partial differential equations of this type.

2.2.3 Energy Equations

Our main emphasis so far has been the derivation of equations for pressure (or hydraulic head) and fluid flow based on mass conservation principles. For natural fluid flows in geological media, these equations are usually more important than equations for energy, which in many cases can be adequately approximated by the initial temperature of the system. However, in engineered systems with significant injection rates, such as CO_2 injection, enhanced oil recovery, or geothermal energy extraction, temperature deviations from the initial state may be significant and of interest. One particular concern arises when temperatures and pressures are near the critical point of the fluid, so that density becomes a strong function of temperature as well as pressure. This requires the consideration of energy balances, which then leads to additional equations that may need to be solved simultaneously with Darcy's law and mass conservation.

Let us denote temperature by the symbol Θ. We begin by noting that fluid density is usually a function of temperature as well as pressure. Therefore, we observe that in the derivation of Equation (2.19), inclusion of the dependence of density on temperature gives an additional term when the time derivative of density is expanded, resulting in the following equation:

$$\rho c_\Sigma \frac{\partial p}{\partial t} + \rho c_{\Sigma,\Theta} \frac{\partial \Theta}{\partial t} - \nabla \cdot \left(\rho \frac{\boldsymbol{k}}{\mu} (\nabla p - \rho \boldsymbol{g}) \right) = \psi,$$

where the total thermal contribution to volume change is defined analogously to the total compressibility,

$$c_{\Sigma,\Theta} \equiv \frac{\phi}{\rho} \frac{\partial \rho}{\partial \Theta} + \frac{\partial \phi}{\partial \Theta}.$$

In order to close our system with this extra temperature dependence, we need to consider additional equations. Our starting point is the principle of energy conservation. In the same way that we work with pressure instead of density as an independent variable for the fluid, it is more natural to work with enthalpy (which is conserved for isobaric processes) instead of energy (which is conserved for isovolumetric processes) when considering the energy balance equation for the fluid phase. For the solid, this is less of an issue, and we will for simplicity of notation use enthalpy also for the solid. We proceed by denoting the specific enthalpy, defined as enthalpy per mass, as $\hbar$, with subscripts f and s indicating fluid and solid. Enthalpy is transported with the fluid, while also dispersing according to Fourier's law. If we denote the Fourier flux as $\boldsymbol{j}_e$, we can express conservation of enthalpy of the combined fluid and solid using the general balance equation, Equation (2.12), with the variables defined as

$$\mathscr{m} = \hbar_f \phi \rho_f + \hbar_s (1-\phi) \rho_s, \quad \mathscr{f} = \hbar_f \rho_f \boldsymbol{u} + \boldsymbol{j}_{e,f} + \boldsymbol{j}_{e,s}, \quad \text{and} \quad \mathscr{r} = \psi_\hbar.$$

Here, $\psi_\hbar$ is a source of energy per volume per time. For completeness we recall that the Fourier heat flux is analogous to Darcy's law, and is given in terms of the heat conductivity $\mathcal{K}$ and the temperature gradient as

$$\boldsymbol{j}_{e,\beta} = -\mathcal{K}_\beta \nabla \Theta_\beta.$$

Here we have used β to denote either the solid (s) or fluid (f). The heat conductivity used here implicitly accounts for the tortuous paths of the porous medium, and therefore does not equal the heat conductivity of the pure substances.

In order to describe energy transfer, we have introduced four new scalar variables: the enthalpies and the temperatures for both solid and fluid. Further, we have introduced two new vector variables (the energy fluxes). However, we have only given one scalar and two vector equations—conservation of energy and Fourier's Law. The remaining equations are provided by three material-dependent scalar equations that complement the conservation equation, which we often refer to as *constitutive equations*. Two such relations are given by the equations of state for the fluid and solid. More specifically, we assume that the fluid temperature is a known function of pressure and enthalpy, and that the solid temperature is a function of enthalpy only (we neglect the weak dependence on solid stresses):

$$\Theta_f = \Theta_f(\hbar_f, p) \text{ and } \Theta_s = \Theta_s(\hbar_s). \tag{2.22}$$

Note that the temperature is an increasing function of enthalpy, so that we can invert the function and therefore can write $\hbar_s = \hbar_s(\Theta_s)$.

There are several approaches to closing the system with a third equation. The simplest approach, which we will follow in this section, is to assume that the fluid and solid temperatures are equal inside the REV, thus $\Theta_f = \Theta_s$. This will be appropriate for relatively homogeneous media, where the spatial scales are short compared to the temporal scales. More complex approaches, which may be needed in fractured or highly heterogeneous formations, involve writing separate balance equations for the fluid and solid, and seeking ways to model the dissipation of heat between the two temperatures.

Using the closure assumption that the temperatures are equal, we can use Equations (2.22) to eliminate the solid enthalpy altogether, since $\hbar_s = \hbar_s(\Theta_s) = \hbar_s(\Theta_f(\hbar_f, p))$. Similarly, we do not have to consider a separate conductivity for the fluid and solid, and we introduce $\mathcal{K}_\Sigma = \mathcal{K}_f + \mathcal{K}_s$. Proceeding as in Section 2.2.1, we can now write the differential form of the energy conservation equation with only $\hbar_f$ as independent variable:

$$\frac{\partial}{\partial t}[\hbar_f \phi \rho_f + \hbar_s(\Theta_f(\hbar_f, p))(1-\phi)\rho_s] + \nabla \cdot (\hbar_f \rho_f \boldsymbol{u} - \mathcal{K}_\Sigma \nabla \Theta_f(\hbar_f, p)) = \psi_\hbar. \tag{2.23}$$

This gives us a fairly general equation for conservation of energy, written in terms of enthalpy. Note that use of enthalpy instead of temperature is important when considering processes involving phase changes of a fluid. Such applications are common, in particular, in oil and gas recovery and geothermal energy extraction. For the purposes of CO_2 storage, phase changes are unlikely to occur within a properly sited aquifer, and are primarily a concern only when considering leakage pathways. However, Equation (2.23) is quite complex, in that it involves two nonlinear equations of state (both for the fluid and solid). Returning to our perspective that the energy equation is of secondary importance to the flow equation for many applications, it is therefore of interest to look at simplifications.

For small variations in temperature and enthalpy, we can linearize equation (2.23). In the linearized equation, when phase change is not involved, it is common to use temperature as the independent variable. We assume that the equation of state is locally invertible to write $\hbar_f = \hbar_f(\Theta_f, p)$. This leads to

$$c_{th,\Theta}\frac{\partial \Theta}{\partial t} + \nabla \cdot (\hbar_f(\Theta, p)\rho_f \boldsymbol{u} - \mathcal{K}_\Sigma \nabla \Theta) = \psi_\hbar - c_{th,p}\frac{\partial p}{\partial t} \tag{2.24}$$

with the effective total specific heat capacity is defined as

$$c_{th,\Theta} \equiv \frac{\partial}{\partial \Theta}[\hbar_f \phi \rho_f + \hbar_s(1-\phi)\rho_s]$$

and the coefficient of the pressure derivative is given as

$$c_{th,p} = \frac{\partial}{\partial p}(h_f \phi \rho_f) - h_s \frac{\partial}{\partial \sigma_{eff}}((1-\phi)\rho_s).$$

Equation (2.24) is the standard heat equation, with the addition of an advective term and a source term due to compression and expansion. Equation (2.24) is widely used and appropriate for a wide range of problems. However, during a phase transition, the definition of heat capacity loses its meaning because the temperature no longer varies with enthalpy. Thus, the linearization leading to Equation (2.24) is no longer valid, and we have to revert to Equation (2.23) for a reliable mathematical model.

2.3 REDUCTION OF DIMENSIONALITY

In many applications, it is convenient to take advantage of the fact that most permeable subsurface formations have horizontal extent that is much greater than the vertical extent. As indicated in Section 1.4, typical horizontal extent of formations is on the order of tens to hundreds of kilometers, while thickness is on the order of tens of, to perhaps one or two hundred, meters. Geometrically, aquifers thus have a comparable aspect ratio to a page of this book. The large disparity in length scales often means that vertical flows are small relative to horizontal flows, and therefore they can be neglected within the formation. This argument is strengthened by the tangent law of flow, which implies that along the boundary separating a formation with higher permeability (aquifer) and a formation with lower permeability (aquitard), the flow in the formation with higher permeability tends toward the direction parallel to the interface while that in the formation with lower permeability tends toward the normal direction relative to the interface (see Box 2.4). When flows in the permeable formation are dominated by the parallel-direction flows, it often makes sense to simplify the governing equations in the aquifer to focus on the horizontal flows only. This leads to a reduction of dimensionality, wherein the three-dimensional descriptions presented in the previous section are replaced by two-dimensional versions of the general flow equations. We can achieve this reduction of dimensionality by integration of the three-dimensional equations over the vertical dimension, thereby eliminating the vertical coordinate and in the process transforming the relevant variables into integrated quantities that are only functions of x_1, x_2, and t.

2.3.1 Vertical Integration

Consider the formation illustrated in Figure 2.3. The permeable formation of interest is bounded above and below by the *T*op and *B*ottom surfaces whose elevations are given by $\zeta_T(x_1, x_2)$ and $\zeta_B(x_1, x_2)$, respectively. This gives a formation thickness of

BOX 2.4 *The Tangent Law*

When a distinct boundary exists between two different geological formations, we refer to the interface between the formations as a "material interface." Such interfaces often occur in layered formations. At this interface, two conditions must apply. The first is that the fluid pressure across the interface must be continuous, which means that the pressure on each side of the interface must be the same as the interface is approached from each side. The second condition is that the mass flux across the interface must be equal, which means that the mass leaving one formation through the interface must be the same as the mass that arrives in the adjacent formation. This requires continuity of the normal flux vector. These two conditions can be combined to examine flows adjacent to a material interface when one of the materials has a hydraulic conductivity significantly different than the other.

Consider the situation shown in Figure 2.B, where a higher permeability material (Material 1) is adjacent to a lower permeability material (Material 2), separated by the interface ζ. We will for simplicity take both materials to be isotropic.

Continuity of normal flux requires that $u_{\perp,1} = u_{\perp,2}$, while continuity of pressure along the interface means implies continuity of hydraulic head which means that $\partial h_1/\partial\zeta = \partial h_2/\partial\zeta$. Noting that $u_{\parallel,1} = -\kappa_1(\partial h_1/\partial\zeta)$ and $u_{\parallel,2} = -\kappa_2(\partial h_2/\partial\zeta)$, the tangents of the angles θ_1 and θ_2 can be related as follows:

$$\frac{\tan\theta_2}{\tan\theta_1} = \frac{u_{\parallel,2}/u_{\perp,2}}{u_{\parallel,1}/u_{\perp,1}} = \frac{\kappa_2}{\kappa_1}.$$

Notice that if $\kappa_1 \gg \kappa_2$, then $\theta_1 \gg \theta_2$, which means that the flow direction in the higher permeability layer is close to parallel to the interface while the direction in the lower permeability zone is close to perpendicular to the interface. This is consistent with the usual assumption of horizontal flow in aquifers and vertical flow in aquitards, assuming the layering (i.e., the interface) is in the horizontal direction.

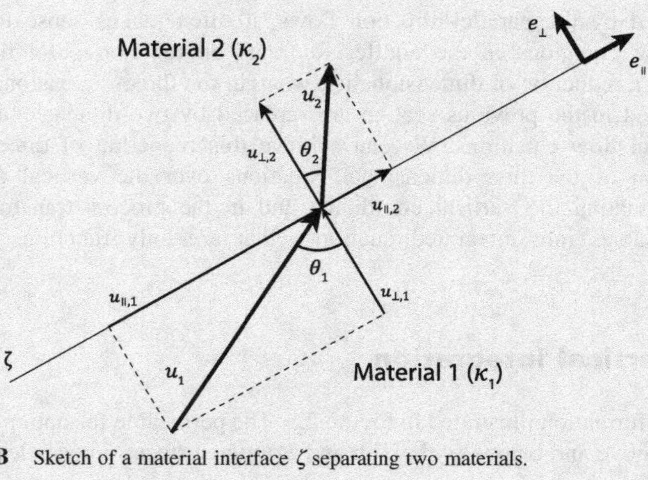

Figure 2.B Sketch of a material interface ζ separating two materials.

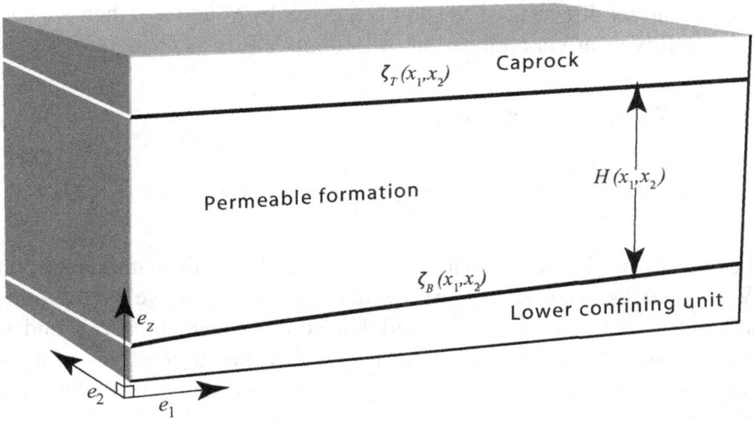

Figure 2.3 Schematic of an aquifer and the notation used to describe the geometry.

$H(x_1, x_2) \equiv \zeta_T(x_1, x_2) - \zeta_B(x_1, x_2)$. We assume the top and bottom boundaries do not differ too much from horizontal in their orientation, although they can vary in space as indicated. The vertical integration procedure simply takes the three-dimensional governing equation, for example, Equation (2.21), and integrates it with respect to the vertical coordinate, between the limits of ζ_B and ζ_T. For simplicity, we begin with a modified form of Equation (2.21) in which Darcy's equation has not been used to substitute for the flow vector. That is,

$$\int_{\zeta_B}^{\zeta_T} \left(S_s \frac{\partial h}{\partial t} + \nabla \cdot \boldsymbol{u} \right) dz = \int_{\zeta_B}^{\zeta_T} \upsilon dz. \tag{2.25}$$

The idea of vertical integration is to derive an equation in which the primary unknowns are the vertically integrated versions of the three-dimensional variables h and $\boldsymbol{u}$. Therefore, we are motivated to exchange the integration and differentiation in Equation (2.25). If we consider first the time derivative term, we can observe that the limits of integration are not dependent on time, and therefore the exchange of differentiation and integration is simple. If we assume that the specific storativity does not change as a function of vertical location, then we can write the following equality,

$$\int_{\zeta_B}^{\zeta_T} S_s \frac{\partial h}{\partial t} dz = S_s \frac{\partial}{\partial t} \int_{\zeta_B}^{\zeta_T} h dz. \tag{2.26}$$

The spatial derivatives require use of Leibnitz's rule (see the Appendix) to exchange integration and differentiation, because the limits of integration are functions of x_1 and x_2. If we denote the components of the flow vector in the x_1 and x_2 directions with the two-dimensional vector $\boldsymbol{u}_\parallel = (u_1, u_2)$, where the "parallel to" subscript notation implies flow in the directions orthogonal to the averaging direction (i.e., parallel to the general lateral direction of the aquifer, given by x_1 and x_2), then we can write the integrated flux term via application of Leibnitz's rule as follows,

$$\begin{aligned}\int_{\zeta_B}^{\zeta_T} \nabla \cdot \boldsymbol{u} dz &= \int_{\zeta_B}^{\zeta_T} \left(\nabla_{\|} \cdot \boldsymbol{u}_{\|} + \frac{\partial}{\partial z} u_z \right) dz \\ &= \nabla_{\|} \cdot \int_{\zeta_B}^{\zeta_T} \boldsymbol{u}_{\|} dz + \boldsymbol{u}_T \cdot (\boldsymbol{e}_z - \nabla_{\|} \zeta_T) - \boldsymbol{u}_B \cdot (\boldsymbol{e}_z - \nabla_{\|} \zeta_B),\end{aligned} \tag{2.27}$$

where the "parallel to" subscript on the gradient operator has the natural interpretation, that is, $\nabla_{\|} = (\partial/\partial x_1)\boldsymbol{e}_1 + (\partial/\partial x_2)\boldsymbol{e}_2$. In writing Equation (2.27), we have assumed that the coordinates z and x_3 are identical (and therefore perpendicular to x_1 and x_2), and used the notation that subscript 'T' or 'B' for the flow vector $\boldsymbol{u}$, or any of its components, implies evaluation at the top ($z = \zeta_T$) or bottom ($z = \zeta_B$), of the formation, respectively.

Note that in Equation (2.27) the flow vector $\boldsymbol{u}$, which is defined point-wise in three dimensions, is converted into a vertically integrated quantity that is now a function of only two space dimensions, x_1 and x_2. We denote this new, vertically integrated variable using the corresponding uppercase letter, so that

$$\boldsymbol{U}(x_1, x_2, t) \equiv \int_{\zeta_B}^{\zeta_T} \boldsymbol{u}_{\|} dz.$$

The remaining two terms in Equation (2.27) correspond to fluxes entering or leaving the formation through the top and bottom boundaries. The vertical dimension has been eliminated via integration, so these fluxes, which are necessary for a proper mass balance, now appear in the governing equation instead of appearing as boundary fluxes (boundary conditions) in the three-dimensional representation. We will denote these fluxes appearing in Equation (2.27) as Υ_T and Υ_B, respectively, and note that in the integrated model, they take the role of source and sink terms.

Let us now return to Equation (2.26). Consider definition of an average value of hydraulic head along the vertical, over the formation thickness H, defined by

$$\bar{h}(x_1, x_2, t) \equiv \frac{1}{H} \int_{\zeta_B}^{\zeta_T} h(x_1, x_2, z, t) dz. \tag{2.28}$$

In Equation (2.28), the overbar notation implies an averaged quantity. We can use this definition in Equation (2.26) to replace the integral with the average,

$$S_s \frac{\partial}{\partial t} \int_{\zeta_B}^{\zeta_T} h dz = S_s H \frac{\partial \bar{h}}{\partial t}. \tag{2.29}$$

Examination of Equation (2.29) reveals that a new coefficient has been created, defined by the product of S_s and H. In the groundwater hydrology literature this new coefficient is called *the storage coefficient*, and is denoted by S. While the specific storativity has dimensions of $[L^{-1}]$, the storage coefficient is dimensionless.

With these integrated equations, we can combine Equations (2.27) to (2.29) to yield the following expression:

$$S\frac{\partial \bar{h}}{\partial t} + \nabla_{\|} \cdot \boldsymbol{U} + \Upsilon_T - \Upsilon_B = \int_{\zeta_B}^{\zeta_T} \upsilon dz \equiv \Upsilon. \tag{2.30}$$

Again we have used an uppercase variable to denote vertically integrated quantities, in this case introducing Υ to represent the vertically integrated volumetric source or sink term.

Equation (2.30) may be seen as a vertically integrated version of the mass balance equation, with compressibility included. We now have the task to include the information contained in the Darcy equation, and to use that to replace the integrated flow vector with either vertically averaged hydraulic head, or some measure of pressure. Let us continue to use the hydraulic head, to be consistent with the compressibility treatment. The procedure for the Darcy equation is analogous to what has already been done: apply vertical integration over the formation thickness to Darcy's law, Equation (2.8), using Leibnitz's rule as appropriate. In most porous media, the layered structure will be roughly horizontal, such that one of the main anisotropy directions of hydraulic conductivity aligns with the vertical direction. We use this assumption and write the conductivity tensor as

$$\boldsymbol{\kappa} = \begin{bmatrix} \kappa_{1,1} & \kappa_{1,2} & 0 \\ \kappa_{2,1} & \kappa_{2,2} & 0 \\ 0 & 0 & \kappa_z \end{bmatrix} = \begin{bmatrix} \boldsymbol{\kappa}_{\|} & 0 \\ 0 & \kappa_z \end{bmatrix}$$

where $\kappa_{\|}$ is the 2×2 horizontal matrix. The assumption that horizontal (parallel) flow directions dominate the system means that vertical flows are insignificant within the formation, which in turn implies very small vertical gradients of hydraulic head (assuming κ_z does not approach zero). Therefore, we can reasonably assume that the hydraulic head within the formation is essentially constant along the vertical direction. If we represent the vertical variation in head as $\tilde{h} \equiv h - \bar{h}$, we can write the following expression,

$$\boldsymbol{U} = \int_{\zeta_B}^{\zeta_T} \boldsymbol{u}_{\|} dz = -\int_{\zeta_B}^{\zeta_T} \boldsymbol{\kappa}_{\|} \nabla_{\|} h dz = -H\overline{\boldsymbol{\kappa}_{\|}} \nabla_{\|} \bar{h} - \int_{\zeta_B}^{\zeta_T} \boldsymbol{\kappa}_{\|} \nabla_{\|} \tilde{h} dz \approx -H\overline{\boldsymbol{\kappa}_{\|}} \nabla_{\|} \bar{h}.$$

Similar to the storage coefficient appearing in Equation (2.29), we define a new parameter associated with the vertically integrated flow terms, called the *transmissivity* $\boldsymbol{T} \equiv H\overline{\boldsymbol{\kappa}_{\|}}$. We note that the arithmetic average of hydraulic conductivity naturally appears in the integrated equation; this is the meaning of the averaged term $\overline{\boldsymbol{\kappa}_{\|}}$.

Combination of the averaged flow equation with the averaged mass balance equation leads to the following equation that governs two-dimensional flow in aquifers,

$$S\frac{\partial \bar{h}}{\partial t} - \nabla_{\|} \cdot \left(\boldsymbol{T} \nabla_{\|} \bar{h}\right) = \Upsilon_{\Sigma}. \tag{2.31}$$

We will refer to Equation (2.31) as the *two-dimensional single-phase flow equation in terms of hydraulic head*, and associate with the Σ subscript the signed sum of the source terms, where the signs are chosen such that the total fluid source in the integrated equation is obtained:

$$\Upsilon_\Sigma = \Upsilon - \Upsilon_T + \Upsilon_B.$$

A similar vertically integrated equation can be derived using pressure instead of hydraulic head as the primary variable. The procedure is identical, involving repeated application of Leibnitz's rule and invocation of vertical equilibrium to simplify the resulting equation. However, there are some differences in approach because hydraulic head and pressure have different functional forms and vary differently as a function of z. One of the advantages of working with hydraulic head is that it is a direct measure of energy or potential, and when a fluid is in vertical equilibrium, the hydraulic head will be constant with respect to vertical location. This allows for direct replacement of h by $\overline{h}$, although we could have chosen any other point-wise value for hydraulic head along the vertical direction, for example the head at the top (h_T) or bottom (h_B) of the formation. When the equation is written in terms of pressure, then vertical equilibrium means that pressure varies linearly with z, and therefore the relationships among $\overline{p}$, p_B, and p_T are not as simple as those for hydraulic head. In this case, we usually find that the choice of either p_B or p_T as the primary unknown is more convenient than the average pressure $\overline{p}$.

The expression for vertical equilibrium in terms of pressure, given a domain that extends from the bottom boundary ζ_B to the top boundary ζ_T, where density is held constant, may be written as follows,

$$p(x_1, x_2, z, t) = p_B(x_1, x_2, t) - \rho g[z - \zeta_B(x_1, x_2)]. \tag{2.32}$$

From Equation (2.32) we see that $p_T = p_B - \rho gH$. Application of vertical integration, with Leibnitz's rule as needed, yields the following expressions when the vertical equilibrium assumption of Equation (2.32) is applied and the permeability matrix has primary anisotropy corresponding to the horizontal and vertical directions (recall that we neglect vertical variation in specific storativity and compressibility):

$$\begin{aligned} \int_{\zeta_B}^{\zeta_T} c_\Sigma \frac{\partial p}{\partial t} dz &= c_\Sigma H \frac{\partial p_B}{\partial t} \\ \boldsymbol{U} = \int_{\zeta_B}^{\zeta_T} \boldsymbol{u}_\| dz &= -\frac{\overline{\boldsymbol{k}_\|}}{\mu} H[\nabla_\| p_B + \rho g \nabla_\| \zeta_B]. \end{aligned} \tag{2.33}$$

When Equations (2.33) are combined with the vertically integrated mass balance equation, we obtain the following equation,

$$C_p \frac{\partial p_B}{\partial t} - \nabla \cdot [\mu^{-1} \boldsymbol{K} (\nabla p_B + \rho g \nabla \zeta_B)] = \Upsilon_\Sigma. \tag{2.34}$$

In Equation (2.34) we have introduced the vertically integrated compressibility coefficient, $C_p = c_\Sigma H$ and the vertically integrated permeability matrix, $\boldsymbol{K} = \int_{\zeta_B}^{\zeta_T} \boldsymbol{k}_\| dz = \overline{\boldsymbol{k}_\|} H$ ·

2.3.2 Simplifications

Here we provide a brief review of a few common simplifications of Equation (2.31), with some comments about their physical importance.

A. Isotropic and Homogeneous Formation

For an isotropic formation, the transmissivity matrix is diagonal and both diagonal entries are the same. This means that a scalar value can be used to replace the matrix. The assumption of homogeneity means that the transmissivity is not a function of space and can therefore be taken outside the spatial derivatives associated with the gradient operator. We then get the following simplification,

$$\nabla_{\parallel} \cdot (T\nabla_{\parallel}\overline{h}) \rightarrow T\nabla^2\overline{h}$$

where ∇^2 is the Laplacian operator, defined by $\nabla^2 = \nabla \cdot \nabla$. Note that the applicability of these assumptions is highly dependent on the geological characterization of the formation and the scale of resolution.

B. Confined Formation, No Sources or Sinks

A confined formation is one in which there are no top or bottom fluxes into or out of the formation. Therefore, the flux terms associated with the boundary, Υ_T and Υ_B, can be set to zero. Most oil and gas reservoirs are bounded by extremely low permeable caprocks, justifying the assumption of a confined formation. If injection and pumping wells are considered as boundary conditions to the porous medium, it is further appropriate to consider that there are no external sources or sinks in the domain, so that

$$\Upsilon_\Sigma \rightarrow 0.$$

C. Steady State, or Incompressible, Conditions

The concept of steady state means that the system is not changing with time. Therefore, the time derivative term can be set to zero. Incompressibility means that neither the fluid nor the solid change volume as a function of pressure changes, implying that the storage coefficient (and the related compressibility coefficients) are all set to zero. Either one of these conditions will eliminate the time-derivative term, such that

$$S\frac{\partial \overline{h}}{\partial t} \rightarrow 0.$$

Multiple Simplifications

Often, it is appropriate to invoke several of the simplifications above. This leads to well known equations from mathematical physics, as we see in Table 2.2.

Table 2.2 Simplifications Applied to Vertically Integrated Models

Simplifications	Equation	Name
A + B	$S\frac{\partial \bar{h}}{\partial t} - T\nabla^2 \bar{h} = 0$	Heat equation (parabolic)
B + C	$\nabla_{\parallel} \cdot (\boldsymbol{T}\nabla_{\parallel} \bar{h}) = 0$	Elliptic equation
C + A	$-\nabla^2 \bar{h} = T^{-1}\Upsilon_{\Sigma}$	Poisson equation
A + B + C	$\nabla^2 \bar{h} = 0$	Laplace equation

2.3.3 Unconfined Formation

Unconfined formations, also referred to as "water-table aquifers," have as their upper boundary a transition zone where below this zone all of the pore space is filled with the fluid of interest (water) and above it the pores are partially or fully filled with one or more other fluids. In groundwater hydrology, this transition zone usually corresponds to what is identified as the *water table*, defined as the surface along which the water pressure is equal to atmospheric pressure, which is equal to zero gage pressure. For unconfined aquifers, the water table is often taken as the top boundary of the domain, with a concomitant assumption that both air and water will exist in the pore spaces above this boundary. While the actual distribution of water and air in the pore spaces above the water table is fairly complex (we will return to this distribution in the next chapters), for now we will assume this transition can be represented by a simple surface that corresponds to a "sharp interface" that separates the zone of full saturation from the zone above it. In the zone above it, we will assume that water only partially fills the pore space, with the remainder of the pore space filled with air. The fraction of the total volume filled with air will be called the "drainable porosity," and denoted by ϕ_d.

Consider a simple unconfined formation where the bottom boundary is horizontal and impermeable, and the top boundary is the water table. We note two important characteristics of this system. First, at the water table, by definition the water pressure is equal to atmospheric pressure, which we set to zero (i.e., we use gage pressure). Because $h = (p/\rho g) + z$, the hydraulic head at the water table will simply equal the elevation of the water table. Because the water table defines the top boundary of our domain, we have $h_T = \zeta_T$. Second, because the top boundary depends on the hydraulic head, the thickness will also depend on the hydraulic head, and we cannot directly apply the results of Section 2.3.1, where we only considered stationary boundaries. Now, both the transmissivity and the storage coefficient will depend on the hydraulic head. Because the hydraulic head is the primary unknown in the equation, the resulting governing equation is nonlinear.

These characteristics can be incorporated in the general framework for vertical integration. In that regard, we may also continue to postulate a system in which the flow is essentially horizontal, and for which vertical equilibrium can be assumed. Such an assumption continues to imply equivalence of the different measures of hydraulic head ($\bar{h} = h_T = h_B$), and now we add the equality between h_T and ζ_T, yielding the following, $\bar{h} = h_B = h_T = \zeta_T$. With the assumption of vertical equilibrium, the integration proceeds exactly as it did for the earlier case. However, we must reevaluate the interpretation of the terms in Equation (2.31).

The normal flux through a moving interface, u_t, can be related to the normal velocity of the interface v_{int} through the following equation,

$$\frac{u_t}{\phi_d} = \frac{\boldsymbol{u}_T \cdot \boldsymbol{v}_n}{\phi_d} - v_{int}.$$

Here, $\boldsymbol{v}_n$ is the vector normal to the interface, $\boldsymbol{u}_T$ is the volumetric flux vector evaluated at the top of the formation (i.e., at the water table), and the drainable porosity is used because it represents the fractional change of pore space occupied by water when crossing the interface. We also use the definition of the normal vector

$$\boldsymbol{v}_n = \frac{\boldsymbol{e}_z - \nabla_{\parallel}\zeta_T}{|\boldsymbol{e}_z - \nabla_{\parallel}\zeta_T|}$$

and the observation that $\boldsymbol{v}_n \cdot \boldsymbol{e}_z = 1/|\boldsymbol{e}_z - \nabla_{\parallel}\zeta_T|$, along with the fact that the interface velocity corresponds to the water table velocity whose rate of change of vertical position, $\partial\zeta_T/\partial t$, can be related to the velocity in the normal direction by the directional vectors. All of this information can then be combined to relate the normal velocity to the vertical movement of the interface as follows,

$$\frac{u_t}{\phi_d} = \frac{\boldsymbol{u}_T \cdot (\boldsymbol{e}_z - \nabla_{\parallel}\zeta_T)}{\phi_d |\boldsymbol{e}_z - \nabla_{\parallel}\zeta_T|} - \frac{\partial\zeta_T}{\partial t}\boldsymbol{v}_n \cdot \boldsymbol{e}_z.$$

We recognize the expression for Υ_T in this equation, and we note that we can rewrite this equation in the following form,

$$\Upsilon_T = \boldsymbol{u}_T \cdot (\boldsymbol{e}_z - \nabla_{\parallel}\zeta_T) = \frac{u_t}{\boldsymbol{v}_n \cdot \boldsymbol{e}_z} + \phi_d \frac{\partial\zeta_T}{\partial t}. \quad (2.35)$$

We also note that at the water table interface, flow through the interface is usually due to infiltration in response to precipitation which we represent by the vertical volumetric flux given by Υ_I (where Υ_I is negative for infiltration due to the direction of the vertical unit vector). Thus, the total flux is related to precipitation through $u_t = (\boldsymbol{v}_n \cdot \boldsymbol{e}_z)\Upsilon_I$, and from this we see that the flux through the top surface can be written in terms of infiltration as

$$\Upsilon_T = \Upsilon_I + \phi_d \frac{\partial\zeta_T}{\partial t}.$$

Substitution of this expression into Equation (2.31), with the accompanying substitution of $\zeta_T = \overline{h}$, leads to the following equation,

$$(\phi_d + S)\frac{\partial \overline{h}}{\partial t} - \boldsymbol{\nabla}_{\parallel} \cdot (\boldsymbol{T}\boldsymbol{\nabla}_{\parallel}\overline{h}) = \Upsilon - \Upsilon_I + \Upsilon_B.$$

We recall that the coefficients S and $\boldsymbol{T}$ are functions of the thickness of the aquifer, and thus the location of the top boundary. For the unconfined case, they therefore depend on the hydraulic head. If the datum of $z = 0$ is taken to coincide with ζ_B, then the thickness is simply equal to $\overline{h}$ and the coefficients are defined by

$$S = S_s\overline{h} \quad \text{and} \quad \boldsymbol{T} = \overline{\boldsymbol{\kappa}}\overline{h}.$$

The drainable porosity, ϕ_d, is often referred to as the *specific yield*, which is denoted by S_y. With these expressions, and an assumption that the bottom flux is zero ($\Upsilon_B = 0$), we obtain

$$(S_y + S_s\overline{h})\frac{\partial \overline{h}}{\partial t} - \boldsymbol{\nabla}_{\parallel} \cdot (\overline{\boldsymbol{\kappa}}\overline{h}\boldsymbol{\nabla}_{\parallel}\overline{h}) = \Upsilon - \Upsilon_I. \tag{2.36}$$

We make three comments about this model. First, for unconfined aquifers, storativity is often much smaller than the drainable porosity, so that the coefficient in front of the temporal derivative is often dominated by the specific yield, S_y. Second, there is an apparent degeneracy of the second-order differential term as $\overline{h} \to 0$. While this is a rather intricate mathematical issue, it turns out to yield few practical problems if approximations to the second-order derivative use the identity $2\overline{h}\boldsymbol{\nabla}\overline{h} = \boldsymbol{\nabla}_{\parallel}\overline{h}^2$. Finally, we remind the reader that the right hand term $-\Upsilon_I$ in Equation (2.36) is positive when infiltration adds water to the subsurface system.

2.4 WELL HYDRAULICS

Well hydraulics is important in most subsurface systems because wells allow subsurface fluids to be extracted, including hydrocarbons from deep formations and groundwater from shallower formations. Wells also allow fluids to be injected from the surface to the subsurface, a process particularly important for carbon capture and storage (CCS) operations. In this section we focus only on single-phase flow, and present several useful solutions associated with groundwater hydrology.

A well is simply a pipe that is placed within a hole drilled into the ground. Most of the pipe is solid, but the portion of the pipe along which fluid will be allowed to pass between the formation and the inside of the pipe will have many small holes in the pipe to allow for hydraulic connection and therefore fluid exchange. The pipe is often referred to as the well casing, and the section with holes is referred to as a slotted section, a screened section, or a perforated section. In general, the deeper the well, the more sophisticated is the design of the "pipe."

Until the recent development of directional drilling techniques, all wells were drilled vertically. As such we will use the vertically integrated hydraulic head formulation of the governing equations. If a well is drilled vertically, and water within the wellbore is pumped to the surface, the head outside the well is reduced and flow develops in the formation radially inward toward the well. Because radial flow is a natural consequence of both pumping and injection wells, it is natural to write the governing equations in a radial coordinate system (r, θ, z). For vertical wells in formations that are isotropic and essentially horizontal, the natural coordinate system is the cylindrical coordinate system with the radial axis ($r = 0$) located along the center line of the well. Note that the equations that govern flow have been written using operators like the gradient, which implicitly adapt to the coordinate system being used—see the Appendix.

Radial flow to or from a well usually means that there are negligible variations in the θ-direction, so that the flow component along the θ-direction, U_θ, can be neglected. For vertically integrated equations, the z-direction has also been eliminated. The resulting equations for confined and unconfined homogeneous formations then take the following forms (note that the assumption of an isotropic integrated permeability and transmissivity is applied),

$$S\frac{\partial h}{\partial t} - T\frac{1}{r}\frac{\partial}{\partial r}\left(r\frac{\partial h}{\partial r}\right) = \Upsilon_\Sigma \tag{2.37}$$

$$(S_y + S_s h)\frac{\partial h}{\partial t} - \frac{1}{r}\frac{\partial}{\partial r}\left(rT(h)\frac{\partial h}{\partial r}\right) = \Upsilon - \Upsilon_I \tag{2.38}$$

Note that we omit the overbars on h for notational simplicity.

The general field of well hydraulics has benefitted greatly from analytical solutions to specific cases of flow to wells. These solutions, developed over the last 100 years, are among the most widely used in the field of subsurface hydrology. They allow for interpretation of field tests involving wells (so-called pumping tests, for example), which are extremely valuable in estimating the key aquifer parameters S and T. They also can be used to make predictions about how a formation will respond to a pumping or injection operation. In the remainder of this section we list several of the most important analytical solutions, chosen both for their widespread use as well as the physical insight they provide, and discuss briefly the implications associated with the solutions. All of the solutions assume that a single vertical well is pumping (or injecting) at a constant rate given by $\mp Q_W$ [L^3T^{-1}] (positive refers to injection). Solutions are expressed as a function of r and t when the problem is transient, and a function of only r when the system is at steady state.

Mathematical representations of wells typically take two different forms. The first recognizes that the well itself is not part of the formation, and therefore the boundary of the domain being modeled is taken to be the radius of the well, r_w. A flux boundary condition is typically imposed at this boundary, corresponding to the well volumetric pumping rate, Q_w,

$$2\pi r_w U_r(r_w) = Q_w. \tag{2.39}$$

The radial flux U_r is of course related to the hydraulic head through the vertically integrated version of Darcy's law. Note that since the geological formation is invariably disturbed by the drilling process, the effective radius of the well may be somewhat greater (or less) than the actual radius of the well casing.

The second option recognizes that the radius of the well, and the corresponding area of the well, is much smaller than the overall size of the domain being modeled. With this in mind, we often assume the formation is continuous throughout space, including the well, and represent the well as a source or sink within the domain. Therefore, the extraction or injection is represented by the right-side source or sink term Υ, in Equation (2.37) or (2.38). Note that the dimensions of Υ are volume per area per time, so we can interpret Υ as the pumping rate divided by the cross-sectional (circular) area of the well. It is often convenient to take further advantage of the very small relative size of the well and represent the well as a so-called line source (or line sink), wherein the radius is allowed to shrink to zero. In this case, the term Υ becomes unbounded because we divide by an area that is shrinking to zero. This unbounded behavior turns out to be acceptable, and is usually represented by so-called Dirac delta distributions, which we denote by the symbol δ. In Cartesian coordinates, if the well is located at $(x_1, x_2) = (x_1^{well}, x_2^{well})$, then the right-side source/sink term becomes $\Upsilon = Q_w \delta(x_1 - x_1^{well}) \delta(x_2 - x_2^{well})$. Note that the Dirac delta distribution for a scalar argument has dimension of $[L^{-1}]$. Additional information about Dirac delta distribution may be found in the appendix. In radial coordinates, the Dirac delta distribution is not as convenient, since it appears at the boundary of our domain (at $r = 0$), and thus is not well defined. Therefore, we usually use a flux boundary condition for an infinitely small well radius to represent the source term in radial coordinates,

$$\lim_{r' \to 0} 2\pi r' U_r(r') = Q_w. \tag{2.40}$$

Case 1: Transient Flow in a Confined Infinite Aquifer

Transient flow to a well in a confined aquifer represents the starting point of almost all well-testing. In particular, it forms the initial assumption from which a pumping test is analyzed in order to try to determine the properties of the formation on a near-well spatial scale. Pumping test are among the most important field measurement techniques, and they remain equally important in CO_2 storage as in other subsurface applications.

In a horizontal, isotropic, homogeneous aquifer that can be considered to be infinite in areal extent, with no leakage along the top or bottom boundaries, and with no other source or sink terms (except a pumping well), Equation (2.37) can be solved for the change in hydraulic head due to pumping by a well with infinitesimal radius, Equation (2.40), that begins at time $t = 0$ and continues at the constant rate Q_w. The solution is given by the following equation (see Box 2.5),

$$h(r,t) - h_{init} = \frac{Q_w}{4\pi T} \int_{\chi}^{+\infty} \frac{e^{-y}}{y} dy = \frac{Q_w}{4\pi T} W(\chi) \tag{2.41}$$

as can be verified by substitution of Equation (2.41) back into the governing equation. In Equation (2.41), χ is a dimensionless group defined by

BOX 2.5 *Derivation of Solution to Transient Flow Equation*

An instructive way to derive Equation (2.41) is by using the concept of self-similar variables to simplify the governing equations. We will review this derivation here. Self-similarity is an important tool when investigating physical phenomena, and we will return to in Section 5.1.1.

The hypothesis of self-similarity is that the solution of $h(r, t)$ can be expressed as a function of a single variable $h(r, t) = h(\chi)$, where the variable $\chi = \chi(r, t)$. By dimensional analysis, it can be shown that χ will always be a product of powers of its arguments, and we therefore write $\chi = br^{a_r}t^{a_t}$ where the coefficients a_r, a_t and b are to be determined. Based on this form, we can calculate

$$\frac{\partial \chi}{\partial r} = \frac{a_r \chi}{r} \quad \text{and} \quad \frac{\partial \chi}{\partial t} = \frac{a_t \chi}{t}$$

and further

$$\frac{\partial h}{\partial t} = \frac{dh}{d\chi}\frac{\partial \chi}{\partial t} = \frac{a_t \chi}{t}\frac{dh}{d\chi} \quad \text{and} \quad \frac{\partial h}{\partial r} = \frac{dh}{d\chi}\frac{\partial \chi}{\partial r} = \frac{a_r \chi}{r}\frac{dh}{d\chi}.$$

Substituting these relations into Equation (2.37), with $\Upsilon_\Sigma = 0$, we obtain after some rearrangement

$$\frac{S}{T}\frac{a_t}{a_r^2}\frac{r^2}{t}\frac{dh}{d\chi} - \frac{d}{d\chi}\left(\chi \frac{dh}{d\chi}\right) = 0.$$

In order to satisfy the assumption that $h = h(\chi)$, we need to choose the unknown coefficients so that r and t do not appear explicitly in the above equation. This is accomplished by letting, for example, $a_r = 2$ and $a_t = -1$. In addition, if we choose $b = -\frac{S}{T}\frac{a_t}{a_r^2} = \frac{S}{4T}$ we get the particularly simple equation

$$\chi \frac{dh}{d\chi} + \frac{d}{d\chi}\left(\chi \frac{dh}{d\chi}\right) = 0.$$

This (essentially) first-order equation for $\chi dh/d\chi$ has the exponential function as solution, from which the solution given in Equation (2.41) follows by integrating again and honoring the boundary conditions.

The key element in this approach was to reduce a partial differential equation in space and time into an ordinary differential equation in a combined space-time variable. Ordinary differential equations are in general much easier to solve, and the resulting solution is called self-similar. While self-similar solutions only exist for special initial and boundary conditions, they nevertheless represent an important class of problems, and frequently allow us to gain significant insight into the underlying physics.

$$\chi = \frac{Sr^2}{4Tt}$$

and the function $W(\chi)$ is called *the well function*. Equation (2.41) is sometimes referred to as the Theis equation, named after C.V. Theis, whose famous work in the 1930s introduced this solution to the groundwater community. Notice that the solution is a function of the dimensionless group χ, which contains both r and t. This dependence on χ means that the solution will be identical for any pairs of r and t that give the same value for χ, which means that the solution depends only on the ratio of r^2/t. Reduction of the solution to dependence on a single dimensionless variable is an important characteristic of the solution, one that we will see again in Chapter 5.

The well function, which simply denotes the exponential integral written in Equation (2.41), can be expanded in a series representation which takes the following form, where the first term is Euler's constant,

$$W(\chi) = -0.5772 - \ln(\chi) + \chi - \frac{\chi^2}{2 \cdot 2!} + \frac{\chi^3}{3 \cdot 3!} - \frac{\chi^4}{4 \cdot 4!} + \cdots$$

When the dimensionless group χ is sufficiently small, usually taken to be $\chi < 0.01$, the series can be truncated after the first two terms, and the well function then becomes a logarithm. This is usually referred to as the "Cooper–Jacob" approximation and forms the basis for a simple method for parameter estimation. Note that the criterion of $\chi < 0.01$ translates, via the definition of χ, into the constraint that $r^2 < (0.04T/S)t$, or $r < C_1\sqrt{t}$, where $C_1 = 0.2\sqrt{T/S}$ is a constant coefficient. This indicates that the radius within which the Cooper–Jacob approximation is valid expands in proportion to the square root of time.

Case 2: Steady-State Flow in a Confined Finite Aquifer

While steady-state flow in a finite confined aquifer is somewhat unrealistic physically, it does provide important qualitative information about system behavior. This includes a clear demonstration of the logarithmic behavior of the solution near the well, and the interpretation of the Cooper–Jacob approximation as the best steady-state approximation to the transient problem, which provides motivation to analyze the transient problem as a series of successive steady-state problems.

The steady-state governing equation for a confined aquifer takes the following form,

$$-T\frac{1}{r}\frac{\partial}{\partial r}\left(r\frac{\partial h}{\partial r}\right) = 0. \tag{2.42}$$

Again we use Equation (2.39) as the inner boundary condition; however, for a confined aquifer at steady state, we also need a finite outer boundary, which we denote by the outer radius r_{outer}. We assume a boundary condition of fixed head at the outer boundary, so that

$$h\big|_{r=r_{outer}} = h_{outer}.$$

We can integrate Equation (2.42) twice, to obtain the solution satisfying the boundary conditions. The resulting solution is known as the Thiem equation,

$$h - h_{outer} = -\frac{Q_w}{4\pi T}\ln\left(\frac{r}{r_{outer}}\right)^2. \tag{2.43}$$

Note that for this case, the solution is actually independent of the well radius, and therefore the solution given in Equation (2.43) would not change if we substitute boundary condition (2.40) for Equation (2.39).

It is interesting to note that Equation (2.43) is identical, up to the choice r_{outer}, with the Cooper–Jacob expansion of the time-dependent equation given in Case 1. If we find the outer radius where both the left-hand side of Equation (2.43) is zero and the two first terms of the well function $W(\chi)$ equal to zero, we find that they coincide if

$$r_{outer} = \sqrt{t}\sqrt{\frac{4T}{S}}\exp(-0.5772). \tag{2.44}$$

We interpret Equation (2.44) as giving the time-dependent outer radius which leads to the best steady-state approximation to the time-dependent problem. We will use this concept further in Chapter 6.

Case 3: Steady-State Flow in a Leaky Infinite Aquifer

While confined aquifers represent a good approximation for many aquifers, often leakage through the confining (aquitard) layers cannot be ignored. This is particularly true for late time, because the drawdown develops a large areal footprint and, in the limit of very large time, the Theis solution from Case 1 predicts an unbounded drawdown as time goes to infinity. In this section we analyze leaky aquifers, first developing the steady-state solution and then the associated transient solution.

A *leaky aquifer* is defined as one in which either Υ_T or Υ_B is nonzero. If we assume Υ_T is nonzero, then the "confining" formation overlying the aquifer of interest must have some nonzero permeability, and be connected to a permeable formation above that can provide water via leakage through the confining formation. It is often reasonable to represent the flow through the confining layer as being essentially vertical, driven by differences in hydraulic head across the layer (recall the tangent law—see Box 2.4). If h_{above} represents the hydraulic head in the formation above the confining layer, and if the confining layer has hydraulic conductivity κ' and thickness H', then the steady-state flux through the confining formation is given by $\Upsilon_T = -\kappa'((h_{above} - h)/H')$. When this is included in the equation for confined aquifers, Equation (2.37), the resulting governing equation stated on the interval $r \in [r_w, \infty)$ is

$$-T\frac{1}{r}\frac{d}{dr}\left(r\frac{dh}{dr}\right)-\kappa'\frac{h_{above}-h}{H'}=0.$$

This equation can be rewritten in the following form,

$$\frac{1}{r}\frac{d}{dr}\left(r\frac{dh}{dr}\right)+\frac{h_{above}-h}{\ell_\kappa^2}=0. \tag{2.45}$$

The outer boundary condition for these equations is set to $h(\infty) = h_{above}$, while the inner boundary condition is given at a finite well radius, Equation (2.39). In Equation (2.45) we have defined the characteristic length $\ell_\kappa = \sqrt{TH'/\kappa'}$, and we have used total derivatives to denote explicitly the fact that the solution for this steady-state problem is a function of only the radial coordinate r.

The solution for this equation involves Bessel functions; for the case of infinite domains with a fixed value of h_{above}, the resulting expression is

$$h(r)-h_{above}=\frac{Q_w}{2\pi T}\frac{K_0\left(\frac{r}{\ell_\kappa}\right)}{\frac{r_w}{\ell_\kappa}K_1\left(\frac{r_w}{\ell_\kappa}\right)}. \tag{2.46}$$

In Equation (2.46), K_0 is the so-called *modified Bessel function of the second kind, order zero*, K_1 is the *modified Bessel function of the second kind, order one*. The well radius is typically much smaller than ℓ_κ, which means the term $(r_w/\ell_\kappa)K_1(r_w/\ell_\kappa)\approx 1$, which simplifies Equation (2.46). A practical calculation that follows from this solution is that a well pumping in a leaky aquifer will draw 95% of its water from leakage within a radial distance of $4\ell_\kappa$, and 99% of its water within a radius of $5\ell_\kappa$. These distances (either $4\ell_\kappa$ or $5\ell_\kappa$) are often used to define the "radius of influence" of the well. Leakage through the confining layer limits the spatial propagation of the pressure (head) perturbation in space. This kind of leakage may be important for CO_2 injection because it impacts the distance of pressure perturbations and therefore the potential area of review for the injection operation.

Note that Equations (2.45) and (2.46) assume that in the aquifer immediately above the pumping-well aquifer, the head h_{above} is constant. More likely, these two aquifers may be part of an extended layered vertical sequence, like the systems shown in Figures 1.7 and 1.9. In such a case, the pressure changes induced by pumping will lead to pressure changes in the aquifer immediately above, which will in turn propagate into the aquifer above it and so on. That is, in general we may have a number of aquifers and aquitards in the sequence, and therefore we choose to analyze the problem more broadly. If the sequence involves N aquifers separated by $N-1$ aquitards, then for any of the aquifers we would write the governing equation including both top and bottom fluxes. For aquifer l in the sequence, the equation would take a form like the following:

$$-T_l\frac{1}{r}\frac{d}{dr}\left(r\frac{dh_l}{dr}\right)-K'_{l+1/2}\frac{h_{l+1}-h_l}{H'_{l+1/2}}+K'_{l-1/2}\frac{h_l-h_{l-1}}{H'_{l-1/2}}=0.$$

Here, ±1/2 refers to the aquitard immediately above or below. We see that the solution for hydraulic head in aquifer l depends on the solutions in aquifers $l - 1$ and $l + 1$. This system of equations can be solved simultaneously, leading to a more complex solution structure, although the solution continues to involve Bessel functions.

Case 4: Transient Flow in a Leaky Infinite Aquifer

The transient response to pumping from a leaky confined system is often important. While somewhat complex in terms of the mathematical description, it will be useful to have the results from this case when we look at injection into deep formations later in the book. As such, in this section we derive the transient response to pumping in a leaky aquifer.

The governing equation for transient flow in a leaky aquifer as described in Case 3 is given by

$$S\frac{\partial h}{\partial t} - T\frac{1}{r}\frac{\partial}{\partial r}\left(r\frac{\partial h}{\partial r}\right) - \kappa'\frac{h_{above} - h}{H'} = 0.$$

We will consider this equation on the radial domain $r \in (0, \infty)$, with inner boundary condition given by Equation (2.40). We assume that the initial condition and the outer boundary are given by $h(r, 0) = h_{above}$ and $\lim_{r\to\infty} h(r,t) = h_{above}$. We denote the steady-state solution obtained in the previous section as h_0, such that

$$h_0 = h_{above} + \frac{Q_w}{2\pi T}K_0\left(\frac{r}{\ell_\kappa}\right).$$

The deviation from steady state $\tilde{h}(r,t) = h(r,t) - h_0(r)$ must then satisfy

$$S\frac{\partial \tilde{h}}{\partial t} - T\frac{1}{r}\frac{\partial}{\partial r}\left(r\frac{\partial \tilde{h}}{\partial r}\right) + \frac{\kappa'}{H'}\tilde{h} = 0 \tag{2.47}$$

with homogeneous (zero) boundary conditions, and the initial condition

$$\tilde{h}(r,0) = h_{above} - h_0(r) = -\frac{Q_w}{2\pi T}K_0\left(\frac{r}{\ell_\kappa}\right).$$

We can solve Equation (2.47) with these initial and boundary values using separation of variables. We write $\tilde{h}(r,0) = h_r(r)h_t(t)$, which when substituted into Equation (2.47) leads to

$$\frac{S}{Th_t}\frac{\partial h_t}{\partial t} + \ell_\kappa^{-2} = -a^2\ell_\kappa^{-2} = \frac{1}{rh_r}\frac{\partial}{\partial r}\left(r\frac{\partial h_r}{\partial r}\right).$$

Here, a^2 is a positive constant. We can immediately solve for h_t and h_r, recognizing that the latter solves the Bessel equation of the first kind, to obtain

$$h_t = \exp\left(-(a^2+1)\frac{T}{\ell_\kappa^2 S}t\right)$$

and

$$h_r = J_0(ar\ell_\kappa^{-1}).$$

Here, J_0 is the *Bessel function of the first kind, order zero*. The zero-order Bessel function of the second kind also solves the equation; however, it does not satisfy the boundary conditions. Now the solution can be written by integrating over a, using $w(a)$ as a weighting function, which must be chosen to satisfy the initial condition:

$$\tilde{h} = \int_0^\infty w(a\ell_\kappa^{-1}) J_0(ar\ell_\kappa^{-1}) \exp\left(-(a^2+1)\frac{T}{\ell_\kappa^2 S} t\right) da.$$

At the initial condition, $t = 0$, so all the exponential terms are equal to 1, and we obtain the equation

$$\tilde{h}(r,0) = \int_0^\infty w(a\ell_\kappa^{-1}) J_0(ar\ell_\kappa^{-1}) da = -\frac{Q_w}{2\pi T} K_0(r\ell_\kappa^{-1}). \tag{2.48}$$

To determine the coefficients $w(a\ell_\kappa^{-1})$, we use two orthogonality properties of Bessel functions

$$\int_0^\infty rJ_0(ar\ell_\kappa^{-1}) J_0(br) dr = \frac{1}{b}\delta(a\ell_\kappa^{-1} - b) \text{ and } \int_0^\infty rK_0(r\ell_\kappa^{-1}) J_0(br) dr = \frac{\ell_\kappa^2}{1+\ell_\kappa^2 b^2}.$$

Multiplying Equation (2.48) by $rJ_0(br)$ and integrating over r, we now obtain the expression for the weighting function,

$$w(a\ell_\kappa^{-1}) = -\frac{Q_w}{2\pi T}\frac{\ell_\kappa a}{1+a^2}.$$

Recalling that $\tilde{h}$ was the deviation from the steady-state solution, we can sum the two components, and use the definition of χ, to obtain the final expression for the transient behavior of the leaky aquifer solution:

$$h(r,t) - h_{above} = \frac{Q_w}{2\pi T}\left[K_0(r\ell_\kappa^{-1}) - \int_0^\infty \frac{aJ_0(ar\ell_\kappa^{-1})}{1+a^2} \exp\left(-\frac{(a^2+1)(r\ell_\kappa^{-1})^2}{4\chi}\right) da\right]. \tag{2.49}$$

Note that the expression in the brackets is only dependent on the dimensionless groups $r\ell_\kappa^{-1}$ and χ, which allows us to define this as the well function for the transient leaky aquitard case, and thus

$$W_\kappa(\chi, r\ell_\kappa^{-1}) \equiv 2\left[K_0(r\ell_\kappa^{-1}) - \int_0^\infty \frac{aJ_0(ar\ell_\kappa^{-1})}{1+a^2} \exp\left(-\frac{(a^2+1)(r\ell_\kappa^{-1})^2}{4\chi}\right) da\right]$$

and

$$h(r,t) - h_{above} = \frac{Q_w}{4\pi T} W_\kappa(\chi, r\ell_\kappa^{-1}).$$

Given that leakage is proportional to $h - h_{above}$, we can integrate Equation (2.49) over space to see the evolution of leakage through the confining layer with time, denoted Q', which satisfies

$$Q'(t) = Q_w\left[1 - \lim_{r\to\infty}\int_0^\infty \frac{r\ell_\kappa^{-1} J_1(ar\ell_\kappa^{-1})}{1+a^2} \exp\left(-\frac{(a^2+1)T}{\ell_\kappa^2 S} t\right) da\right]. \tag{2.50}$$

Here, J_1 is the *Bessel function of the first kind, order one*. At steady state, when $t \to \infty$, the integral term tends to zero and this equation simplifies to the natural observation that $Q' = Q_w$. We further observe that since Equation (2.50) involves the limit of $r \to \infty$, the integral only contains a single dimensionless group, which means that leakage through the caprock can be written concisely as

$$\frac{Q'(\chi_2)}{Q_w} = 1 - \lim_{r\ell_\kappa^{-1} \to \infty} \int_0^\infty \frac{r\ell_\kappa^{-1} J_1(ar\ell_\kappa^{-1})}{1+a^2} \exp\left(-(a^2+1)\chi_2\right) da$$

where the new dimensionless group χ_2 is defined as

$$\chi_2 = \frac{(r\ell_\kappa^{-1})^2}{4\chi} = \frac{Tt}{\ell_\kappa^2 S}.$$

These equations give the fraction of injected fluid which leaks, as a function of time, given a leaky aquitard. From our earlier remarks in Case 3, when evaluating numerically the limit $r\ell_\kappa^{-1} \to \infty$, it is sufficient to consider values of $r\ell_\kappa^{-1} > 5$ in order to obtain less than 1% error in the approximation. We will return to these expressions in Chapter 6, when we discuss leakage and multi-aquifer systems in more detail.

Case 5: Steady State Flow in a Finite Unconfined Aquifer

Finally, we review unconfined aquifers. They are important in hydrology and water management, but they also impact our analysis of CO_2 storage. However, the CO_2 storage problem is upside-down relative to unconfined aquifers: we are interested in the distribution of the lighter fluid (CO_2), under a caprock, rather than the denser fluid (water) over an impermeable bedrock. As we will see in the next chapters, the equations governing these problems have many similarities, which makes familiarity with the solution to problems in unconfined aquifers useful when studying CO_2 storage.

The equation for steady-state flow in a homogeneous, isotropic, unconfined aquifer that has finite areal extent and no recharge may be written as follows:

$$-\frac{1}{r}\frac{d}{dr}\left(rh\frac{dh}{dr}\right) = 0. \tag{2.51}$$

We consider the same bounded domain and boundary conditions as Case 2, keeping in mind the inner boundary condition must take into account the nonlinear flux expression,

$$U_r\big|_{r=r'} = -\kappa h \frac{dh}{dr}\bigg|_{r=r'}.$$

While Equation (2.51) is nonlinear, which would often imply that it cannot be solved analytically, it has a special form that allows an analytical solution to be derived. When the transmissivity is a linear function of the primary unknown, h, the equation can be rewritten to yield h^2 as the unknown in the equation. That is, Equation (2.51) can be rewritten as follows:

$$-\frac{1}{r}\frac{d}{dr}\left(\frac{r}{2}\frac{dh^2}{dr}\right)=0.$$

This equation can be integrated directly, and the solution is a combination of a constant function and a logarithmic function, which takes the following form,

$$h^2(r)=h_{outer}^2-\frac{Q_w}{\pi\kappa}\ln\left(\frac{r}{r_{outer}}\right). \tag{2.52}$$

The positive root of this equation satisfies the boundary condition. The solution structure given by Equation (2.52) turns out to be typical of problems involving flow to a well in unconfined systems, which is the simplest example of a formation that has an interface separating two distinct regions of fluids. We will see in Chapters 3 and 4 that the flow of multiple fluids, for example, brine and CO_2, may be described by equations similar to Equation (2.51), with solutions derived in Chapter 5 that are similar to Equation (2.52).

Graphical Form of Well Functions

Well functions have typically been applied in inverse modeling, where the pressure response at the pumping (or some monitoring) well is observed over time, and the coefficients of the equations are found by a best-fit algorithm. As such, the radial distance is usually given, and it is the temporal variation that is of interest. We therefore plot the well function W_κ as a function of dimensionless time, for various choices of $r\ell_\kappa^{-1}$ in Figure 2.4.

The solid lines of Figure 2.4 represent the leaky aquifer transient solution (Case 4), with dimensionless groups decreasing from bottom to top with values $r\ell_\kappa^{-1}=(5,4,2,1,0.5,0.1,0.05,0.01,0.001)$, where the smallest value is essentially

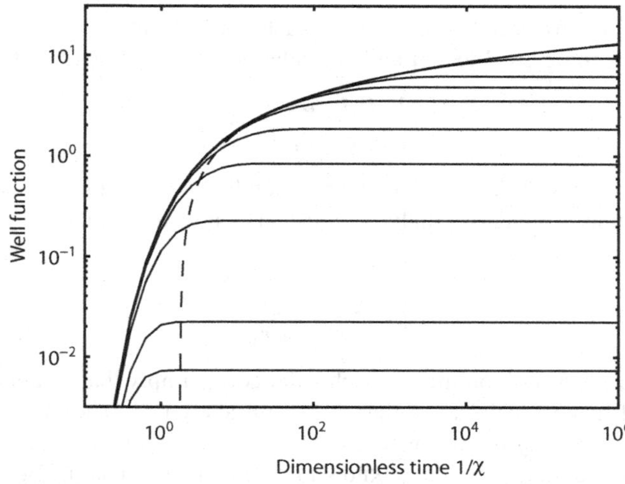

Figure 2.4 Graph of well function for transient leaky aquifers as a function of time.

converged to the Theis solution derived in Case 1 (upper limit of curves). Note that the two lowest curves correspond to $r\ell_\kappa^{-1} = 5$ and 4, and thus correspond to the response at the typical choices of radius of influence discussed in Case 3. Also shown in a dashed line is the Cooper–Jacob approximation to the Theis well function, which corresponds to a particular steady-state solution (Case 2). We see that the Cooper–Jacob approximation is an increasingly accurate approximation for late time (large χ^{-1}), while it is not valid at early times. Indeed, we see from the figure that the Cooper–Jacob approximation only predicts a positive drawdown for $\chi^{-1} > 1.78$. The steady state values of the transient solutions naturally correspond to the steady-state leaky aquifer solution derived in Case 3.

2.5 GEOMECHANICAL RESPONSES

In regions where substantial changes in fluid pressure result as a consequence of injection or pumping, the interactions between the fluids and the rock may need to be taken into account.

The usual approach when considering mechanical deformation in porous media is to use a model of elastic deformation. This model is based on a balance of forces (conservation of momentum), together with models of stress and strain in porous media. For the time scales of interest, it is generally acceptable to neglect acceleration, so that the momentum balance simply relates the body forces (given by gravity), to the shear forces, modeled as the divergence of the stress. The linearized model for strain is related to the gradient of deformation. Finally, Hooke's law gives a linear relationship between stress and strain. This leads to an elliptic system of differential equations, for which much theory has been developed.

Of particular interest for CO_2 storage are possible changes in land surface elevation above an injection operation, and the possible generation or amplification of fractures in the rock. The first is of interest for monitoring considerations, while the second can be critical for enhanced fluid transport and possible leakage. We first consider fractures and then examine the issue of changes in land surface elevation. Almost all geologic media have inherent stresses due to a history of deformation over geologic time scales. The result of large variations in stress, and in particular shear stress, is the possibility of fracturing the rock. Once a fracture exists, it may be open or closed depending on the neighboring stress conditions. Stress changes associated with injection may therefore potentially open existing fractures that initially were closed, or create new fractures through cracking of the porous material.

While modeling of stress fields using continuum mechanics is feasible, modeling the growth and evolution of new and existing fractures remains a challenging problem. Models of flow in porous media that are coupled with an active, fracturing reservoir are usually unable to provide reliable quantitative results. We will therefore focus our attention on practical operational limits to injection that avoid fracture generation. We will also consider simple estimates for ground elevation rise.

2.5.1 Fracture Pressure

Fracturing the porous medium near a well is of great interest to the oil and gas industry, since fractures lead to higher effective permeabilities, and consequently, lower (and cheaper) applied well pressures. Therefore, a certain experience exists regarding injection rates and fracture generation. An estimate of the pressure required to fracture a medium, p_{frac}, can be obtained with the help of the so-called overburden pressure, which is the weight of the whole porous medium. For a homogeneous medium, the overburden pressure is given approximately as

$$p_{ovrb} \approx p_{surf} + \int_{\zeta_{inj}}^{\zeta_T} (\phi\rho_f + (1-\phi)\rho_s)\, g dz'.$$

Here we have used p_{ovrb} to represent the overburden pressure, ζ_{inj} and ζ_T to denote the injection and surface heights such that the integral is over the whole overburden, and ρ_f and ρ_s to represent the fluid and rock densities, respectively. CO_2 injection should always take places below the overburden pressure because pressures above that value will lead to zero or negative effective stress values. This can allow for opening of fractures and faults. The requirement that injection pressure remains below the overburden pressure provides a practical constraint on the injection rate. As an example, current regulations for acid gas storage in Alberta prescribe 90% of the overburden pressure as an upper limit. Note that the overburden pressure is often referred to as the fracture pressure, although fracturing may occur for lower pressures than the overburden pressure depending on the stress state of the geological formation.

The integrand in the expression for overburden pressure, $(\phi\rho_f + (1 - \phi)\rho_s)g$, is often referred to as the fracture gradient. Since porosity and densities vary with depth, the fracture gradient is often not known exactly. One approach to estimate the overburden pressure is to conduct a fracture test, either in the injection formation or in a nearby formation, to estimate the fracture gradient. In any case, one should be aware that the fracture gradient only gives a crude approximation to the actual fracture pressure at any point in a formation. Since fractures are initiated due to local variations in stress, actual fracture pressure within an aquifer will be variable similarly to porosity and permeability.

2.5.2 Ground Elevation Rise

Subsidence is a well-known consequence of both groundwater and hydrocarbon extraction from the subsurface. One of the most famous examples is the case of Mexico City, where subsidence of several meters has led to problems impacting facilities as varied as buildings and sewage systems. The cause of the subsidence is a drop in fluid pressure within unconsolidated lake deposits associated with groundwater withdrawal. As we recall from Section 2.2.2, this implies that a greater fraction of the overburden has to be carried by the effective stress of the solid, thereby leading to compression and compaction.

By analogy to subsidence caused by fluid extraction, injection of large amounts of fluids into the subsurface can be expected to lead to expansion instead of consolidation, thereby leading to a rise of the land surface. This has been observed at the In Salah CO_2 storage site in Algeria, where satellite images indicate a ground-level rise on the order of several centimeters in the vicinity of the injection wells. Without going into the full topic of mechanics and nonlinear deformation, we will look at a simple model for the geomechanical response associated with fluid injection. While this simplified analysis has limited quantitative value, it serves the purpose of giving us an understanding of some of the driving forces of the expected qualitative system behavior.

Assuming that the lateral extent of the formation is constant, the volume change of an aquifer is proportional to a change in its thickness. To be more precise, we use the fact that the mass of solid along any vertical trace is conserved, so that

$$\frac{\partial}{\partial t}\int_{\zeta_B}^{\zeta_T}(1-\phi)\rho_s dz = 0. \tag{2.53}$$

Here, ζ_B and ζ_T refer to a lower, unperturbed (bottom) surface, and the land (top) surface, respectively. Recall the concept of compressibility of the pore space from Section 2.2.2. By combining this concept with the dependence of the solid density on the effective stress, we can define the compressibility of the porous medium with respect to effective stress, and therefore also fluid pressure. We denote this as

$$c_s \equiv \frac{d}{d\sigma_{eff}}\ln((1-\phi)\rho_s) = -\frac{d}{dp}\ln((1-\phi)\rho_s)$$

where we use the logarithm for convenience, as will be seen in the next two equations. We now expand Equation (2.53) in terms of pressure, to obtain

$$\frac{\partial p}{\partial t}\frac{d}{dp}\int_{\zeta_B}^{\zeta_T}(1-\phi)\rho_s dz = 0.$$

Assuming that we are considering problems with pressure change, we can divide by the time rate of change of pressure. The integral term can then be evaluated by Leibnitz's rule, from which we obtain the following expression for the top and bottom surfaces

$$(1-\phi)\rho_s|_{\zeta_T}\frac{d\zeta_T}{dp} - (1-\phi)\rho_s|_{\zeta_B}\frac{d\zeta_B}{dp} = -\int_{\zeta_B}^{\zeta_T}\frac{d}{dp}(1-\phi)\rho_s dz.$$

For the sake of the argument, consider a vertical section that is homogeneous, with a hydrostatic pressure so that the pressure change is also independent of elevation. Then we obtain

$$\frac{1}{H}\frac{dH}{dp} = c_s. \tag{2.54}$$

While the compressibility will, in general, vary with soil or rock type, and therefore with depth, and its value may depend on stress history, Equation (2.54) can still be expected to provide useful order-of-magnitude estimates for practical screening calculations.

For ranges where the compressibility can be considered linear in terms of pressure, we can integrate Equation (2.54) to obtain an expression relating the height of the aquifer to the change in pressure (at some datum) from some reference pressure p_0

$$H(p) - H(p_0) = \exp\left[c_s(p - p_0)\right].$$

Once the deformation of individual formations has been estimated, the sum of the deformations will provide an indication of the expected ground-level disturbance due to injection.

2.6 NUMERICAL SOLUTIONS

We will comment briefly on numerical solutions for the vertically integrated flow equation, for example, Equation (2.31). Numerical approximations may be derived by direct approximation of the derivatives in the equation, or we can return to the original statement of mass balance and write numerical approximations as direct mass balance statements on discrete subdomains of the overall problem domain. We choose the second option, since it allows us to preserve conservation laws in our numerical methods. We begin by writing the mass balance equation, analogous to Equation (2.12), but now for a two-dimensional domain. We assume that the storage coefficient, the transmissivity, the vertically integrated source/sink term, and the top and bottom flux terms are all well defined. Then, over an averaging area in the (x_1, x_2) plane, we write the integral mass balance statement as follows:

$$\int_\omega \frac{\partial \mathscr{m}}{\partial t} dA = -\oint_{\partial\omega} \mathscr{f} \cdot \mathbf{v}_n d\ell + \int_\omega \mathscr{r} dA. \tag{2.55}$$

If the domain of interest is broken into subdomains, each numbered 1 through N, such that the entire domain is covered and no subdomains overlap, then we can apply Equation (2.55) locally over each of the subdomains. The fluxes will be defined along the boundary of the subdomain (denoted by $\partial\omega$) and required to match the fluxes defined in neighbor subdomains which share common boundary segments. While this approach can be quite general, with the subdomains taking on different kinds of shapes, we will illustrate the approach by assuming that all subdomains are rectangular and collectively form a structured rectangular array of grid cells. In this case, we can number the cells along the x_1 and x_2 directions using subscripts i and j, respectively, with N_1 cells along x_1 and N_2 cells along x_2. Then any cell in the interior of the domain will share its boundary with four adjacent cells, as illustrated in Figure 2.5. If single (average) values of the mass density, $\mathscr{m}$ and source $\mathscr{r}$ are used within any cell, and single (average) values of the flux are used along each of the four sides of a given cell, then the discrete version of the mass balance Equation (2.55) may be written as follows:

$$\Delta_{x_1}^{i,j} x_1 \cdot \Delta_{x_2}^{i,j} x_2 \cdot \frac{d \mathscr{m}_{i,j}}{dt} = \Delta_{x_2}^{i,j} x_2 \cdot \Delta_{x_1}^{i,j} \mathscr{f}_1 + \Delta_{x_1}^{i,j} x_1 \cdot \Delta_{x_2}^{i,j} \mathscr{f}_2 + \Delta_{x_1}^{i,j} x_1 \cdot \Delta_{x_2}^{i,j} x_2 \cdot \mathscr{r}_{i,j}. \tag{2.56}$$

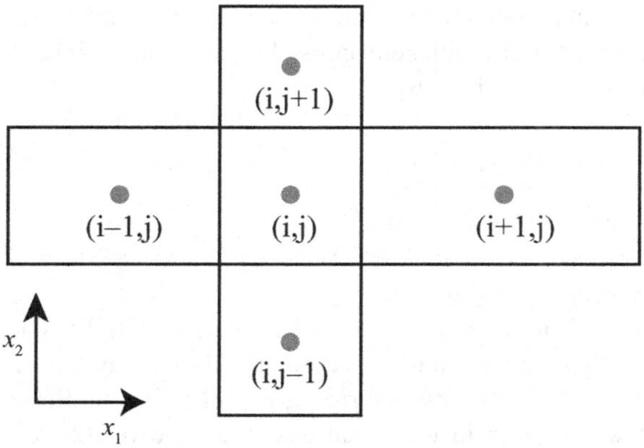

Figure 2.5 Parallelogram grid and indices.

Here, the operator $\Delta_{x_1}^{i,j}$ is the central difference in the x_1 coordinate for the cell (i, j). For a function $u(x_1, x_2)$, the operator is defined as follows:

$$\Delta_{x_1}^{i,j} u(x_1, x_2) \equiv u|_{(i+\frac{1}{2},j)} - u|_{(i-\frac{1}{2},j)}$$

where subscript $i+\frac{1}{2}$ denotes the right boundary of the cell (i, j), while $j-\frac{1}{2}$ would be the front boundary, and so on. Note that this implies that, for example, $\Delta_{x_1}^{i,j} x_1$ is the width of the cell in the x_1 direction. Substitution for the mass density term using

$$\frac{d m_{i,j}}{dt} = S_{i,j} \frac{dh_{i,j}}{dt}$$

along with the following typical expression for the flux terms (assuming that the grid is aligned with the principal axes of anisotropy, so that $T_{1,2} = T_{2,1} = 0$),

$$\not{f}_1 \Big|_{i+\frac{1}{2},j} = -\left(T_{11} \frac{\partial h}{\partial x_1} \right)_{i+\frac{1}{2},j} \tag{2.57}$$

with a similar expression for $\not{f}_2$. The simplest approximation for Equation (2.57) is to take a two-point approximation for the derivative in the equation, and to approximate the transmissivity coefficient at the cell interface as a weighted average of the transmissivity values in the two adjoining cells. If the cells are viewed as being homogeneous, so that the properties within the domain are approximated as piecewise constants (a patchwork of constant values over each of the cells), then flow across the boundary of two cells is analogous to flow perpendicular to a pair of homogeneous layers. Because the effective coefficient for flow perpendicular to layers is given by the harmonic average weighted by the cell lengths (see Box 2.1), this is often taken as the definition of the coefficient evaluated at the cell interface. This then leads to approximations for the flux terms like the following:

$$\not{f}_1 \Big|_{i+\frac{1}{2},j} = -T_{11} \frac{\partial h}{\partial x_1}\Big|_{i+\frac{1}{2},j} \approx -T_{11}^{i+\frac{1}{2},j} \frac{\Delta_{x_1}^{i+1/2,j} h}{\Delta_{x_1}^{i+1/2,j} x_1}.$$

Note that when we apply the difference operator to an interface location, it takes the difference of the centroid locations for the adjacent cells. Further, the weighted harmonic average for the transmissivity is given by

$$T_{11}^{i+\frac{1}{2},j} = \frac{\Delta_{x_1}^{i,j} x_1 + \Delta_{x_1}^{i+1,j} x_1}{\Delta_{x_1}^{i,j} x_1 \cdot \left(T_{11}\big|_{i,j}\right)^{-1} + \Delta_{x_1}^{i+1,j} x_1 \cdot \left(T_{11}\big|_{i+1,j}\right)^{-1}}.$$

Analogous expressions apply to the x_2-direction.

With these coefficients, the resulting equation is an ordinary differential equation in time where the set of spatially discrete unknowns correspond to the hydraulic head values in each of the grid cells. For any interior grid cell, five of these discrete, unknown head values will appear in the equations: the head in the cell of interest plus the heads in the right, left, front, and back neighbor cells. For grid cells that are adjacent to a boundary, the boundary conditions of the problem serve to replace the information that would come from the neighbor cell in the direction of the boundary. For example, if a flux is given along the boundary, then this is included directly into the discrete equation for that cell (the flux term corresponding to the boundary term would correspond to known information in, e.g., Equation (2.56)).

The time dimension is usually discretized as well, often using a relatively simple time-stepping method. The time domain is typically discretized using discrete time levels t^n, denoted by integer counter n and beginning with the initial time, denoted by $n = 0$. At $n = 0$, the solution is known everywhere by the initial condition. This initial information is then propagated to the first time level, t^1, by standard time-stepping such as a weighted Euler approximation, where the time derivative is approximated as $\left(h_{i,j}^1 - h_{i,j}^0\right)/\Delta t^0$ with $\Delta t^0 = t^1 - t^0$. The flux terms are evaluated at a discrete time level between t^0 and t^1. For an explicit calculation, the so-called forward Euler method is used, where all fluxes are evaluated at the previous time level, in this case at t^0. When this is done, each discrete equation contains only one unknown value (i.e., only one value at the new time level), so each equation can be solved *explicitly*. If the flux terms are evaluated at the new time level, t^1, then all five values of the hydraulic head in any given equation will be evaluated at the new (unknown) time level, resulting in the need to solve a set of coupled equations, which means a matrix solution. The solution is therefore only known *implicitly*, and the method is referred to as a backward Euler approximation. In general, a weighted average of the two can be used, with an arbitrary weighting factor that spans the forward and backward approximations. While more computationally intensive per time step, due to the matrix solutions, implicit methods are usually preferred because they are numerically stable for all time-step sizes, whereas explicit methods are only conditionally stable and usually require very small time steps.

The most well-known numerical simulator in groundwater hydrology is MODFLOW, a software package that has been developed by the U.S. Geological Survey. It is based on the kinds of numerical approximations outlined in this section. While the numerical approximations are simple, they are often quite effective in providing insights into overall flow characteristics and highlighting sensitivities to certain flow parameters.

2.7 COMPRESSION AND RECONSTRUCTION

As a final section in this chapter focused on single-phase flow in porous media, we present a particular way to think about some of what has been derived to this point. We first observe that in much of what we do, we represent parameters and processes that occur on a smaller length scale with derived parameters and modified equations on a larger scale. We did this initially when defining variables like porosity that are meant to represent a single bulk measure of the highly complex pore space of a porous medium. We might think of this as taking a large amount of information on the fine scale and compressing it to a much smaller set of information defined on a coarser scale. In the case of single-phase flow in porous media, all of the geometric complexity of the tortuous network of connected pore spaces is compressed into two measures representing the complexity of the overall pore space: porosity and permeability. Other geometric measures might be used in conjunction with the these two—for example, the specific surface area is sometimes used, defined as the total surface area between the solid and fluid (i.e., the surface of the solid) divided by the total volume of porous medium. However, porosity and permeability are often the only explicit variables defined at the coarse scale. We will see when we move on to two-phase flow that we need not only parameters, but also functional relationships (in this case involving the so-called relative permeability and capillary pressure functions) to characterize the geometric information we are interested in at the coarser scale.

Another example of a compression of information is the integration of our three-dimensional equations to produce two-dimensional equations. In the process of this vertical integration, we eliminate one spatial dimension, and thereby eliminate our ability to represent explicitly any variations along that integrated direction. We observe that in the process of vertical integration, the permeability field which is, in general, a function of all three spatial dimensions, must now be represented by vertically integrated quantities that only vary with respect to x_1 and x_2. Vertical integration transforms hydraulic conductivity into transmissivity and specific storativity into the storage coefficient. Symbolically, we may represent *the compression operator* by the letter $\mathcal{C}$. We would then write $T = \mathcal{C}\kappa$ and $S = \mathcal{C}S_s$. In these cases, compression corresponds to vertical integration.

The complement to compression is *reconstruction*. If the compressed information were provided—for example, a value of porosity or transmissivity—how might the original fine-scale variables be reconstructed from this coarse-scale information? Obviously, this is a more difficult problem. If no information is available about the fine scale, then there may be little that can be done. However, we can sometimes make reasonable estimates, and perhaps even better than that. Consider the case of transmissivity, and recall its definition, which is rewritten here,

$$T(x_1, x_2) = \mathcal{C}\kappa \equiv \bar{\kappa}(x_1, x_2)H(x_1, x_2) = \int_{\zeta_B}^{\zeta_T} \kappa(x_1, x_2, z)dz.$$

This equation gives the definition of transmissivity, which is basically the expression for the compression of $\kappa(x_1, x_2, z)$ into $T(x_1, x_2)$. The reconstruction challenge is to take the compressed information, $T(x_1, x_2)$, and reconstruct the more detailed information about $\kappa(x_1, x_2, z)$. That is, if $T = \mathcal{C}\kappa$ defines compression, then we define the *reconstruction operator* analogously as $\hat{\kappa} = \mathcal{R}T$. Note that the hat denotes a reconstructed variable. If the vertical distribution of the hydraulic conductivity is known, then the task is simple because the desired information is already known. However, that is usually not the case. So the reconstruction might involve a combination of information sources, conditioned on the value for transmissivity. For example, if we have a shallow sand and gravel glacial outwash aquifer, statistical characterization from detailed measurements made on other similar formations might be used to guide the reconstruction. The variability of hydraulic conductivity along any given vertical line might be taken to be a correlated random field with an underlying lognormal distribution of values. Using data like that from well-studied field sites like Base Borden in Canada and the Cape Cod site in the United States, one might postulate a vertical correlation length on the order of tens of centimeters, and a variance or standard deviation based on field measurements. With this random structure, the mean of the distribution could be chosen to give the specific value of the transmissivity. These might be further conditioned on specific data from the aquifer under consideration. Or some other means of reconstruction might be employed, in an effort to represent in some reasonable way the fine-scale structure of the medium. For relatively simple single-phase flow, these reconstructions may not be very important. It is for more complex problems, involving more than one fluid phase (e.g., the CO_2 problem—see Chapters 3 and 4, Sections 3.9 and 4.2, respectively) and complex contaminant transport, where the reconstruction may be much more important. For multiphase flow problems it allows phase saturations to be represented on the fine scale, and for component transport it might allow concentration fields to be similarly represented on the fine scale. For the latter, such reconstruction might give hope to solving nonlinear reactive transport problems that resolve important (small) scales of heterogeneity but still allow for most of the computations to be carried out at the coarse scale.

We can perform these kinds of reconstructions for all of the variables of interest, although some are obviously more challenging than others. As a final example, consider porosity, where the reconstruction involves a fine-scale description of the pore space. While direct measurements of the pore space structure are highly unlikely, there are often other sources of information and processes that can give approximate structures. For soils, there are the basic classification methods (sand, silt, loam, etc.), there may be information about grain-size distributions, and there may be reconstructions in the literature for similar kinds of materials, for example, using pore-scale network models, which could provide guidance for the reconstruction. Similarly, for consolidated rocks, significant recent advances have occurred in geometric reconstruction of detailed rock geometry, combining coarse-scale information like porosity and permeability with depositional and consolidation history to create quite realistic rock structures.

All of this provides a general framework within which models for flow in porous media can be constructed. In this chapter, we have concentrated on single-phase flow while introducing ideas that will be carried through all of the remaining chapters. These include the important idea that equations can be written at coarser scales through systematic use of compression operations, with the most important example being vertical integration. Equally importantly, we can use information about finer scales of the problem to define reconstruction operators that allow fine-scale variability to be represented, thereby leading to better resolution of the underlying processes. These key ideas will be carried forward, along with the more basic ideas of Darcy's law, mass conservation, and compressibility. These all form the foundation for our analysis of the CO_2 storage problem.

2.8 FURTHER READING

Groundwater Text Books

BEAR, J. 1988. *Dynamics of Fluids in Porous Media*. Mineola, NY: Dover.

DOMENICO, P.A. and F.W. SCHWARTZ. 1998. *Physical and Chemical Hydrogeology*, Second Edition. New York: John Wiley and Sons.

FAURE, G. 1998. *Principles and Applications of Geochemistry*, Second Edition. Upper Saddle River, NJ: Prentice Hall.

FETTER, C.W. 1994. *Applied Hydrogeology*, Third Edition. New York: Macmillan College Publishing.

FJAR, E., R.M. HOLT, A.M. RAAEN, R. RISNES, and P. HORSRUD. 2008. *Petroleum Related Rock Mechanics*, 2nd Edition. Amsterdam: Elsevier Science.

FREEZE, R.A. and J.A. CHERRY. 1979. *Groundwater*. Englewood Cliffs, NJ: Prentice Hall.

HEATH, M.T. 2002. *Scientific Computing: An Introductory Survey*, Second Edition. Boston: McGraw Hill.

KRUSEMAN, G.P. and N.A. DE RIDDER. 1990. *Analysis and Evaluation of Pumping Test Data*, Second Edition. Wageningen, The Netherlands: ILRI Publication 47, International Institute for Land Reclamation and Improvement.

SPITZ, K. and J. MORENO. 1996. *A Practical Guide to Groundwater and Solute Transport Modeling*. New York: John Wiley and Sons.

WANG, H.F. and M.P. ANDERSON. 1982. *Introduction to Groundwater Modeling: Finite Difference and Finite Element Methods*. New York: W.H. Freeman and Co. (Reprinted by Academic Press, San Diego, CA, 1995).

Papers

HANTUSH, M.S. 1964. Hydraulics of wells. In: *Advances in Hydroscience*, Vol. 1. V.T. Chow, ed. New York: Academic Press, pp. 281–432.

HUBBERT, M.K. 1940. The theory of ground-water motion. *Journal of Geology*, 48, 785.

HUNT, B. 1985. Flow to a well in a multi-aquifer system. *Water Resources Research*, 21(1), 95–100.

LEBLANC, D.R., S.P. GARABEDIAN, K.M. HESS, R.D. QUADRI, K.G. STOLLENWERK, and W.W. WOOD. 1991. Large-scale natural gradient tracer test in sand and gravel, Cape Cod, Massachusetts. 1. Experimental design and observed tracer movement. *Water Resources Research*, 27(5), 895–910.

SUDICKY, E.A. 1986. A natural gradient experiment on solute transport in a sand aquifer—Spatial variability of hydraulic conductivity and its role in the dispersion process. *Water Resources Research*, 22(13), 2069–2082.

THEIS, C.V. 1935. The relationship between the lowering of the piezometric surface and the rate and duration of discharge of a well using ground-water storage. *American Geophysical Union Transactions*, 16, 519–524.

Other Materials

HARBAUGH, A.W. 2005. MODFLOW-2005, the U.S. geological survey modular ground-water model—The ground-water flow process: U.S. geological survey techniques and methods 6-A16.

Chapter 3

Two-Phase Flow in Porous Media

This chapter builds on the material in Chapter 2 to consider porous media systems in which multiple fluid phases are present in the pore space. This is necessary to analyze the geological storage system because when CO_2 is injected into a subsurface formation, it will initially exist as a separate fluid phase. The presence of multiple fluid phases implies a much more complex system within the pore space of the medium.

When CO_2 is injected into a brine-filled formation, it will be injected at a sufficiently high pressure to displace much of the brine in the pore spaces, in the process creating fluid–fluid interfaces at the pore scale that allow the two fluids to coexist in the pore spaces. Those pore-scale fluid–fluid interfaces play important roles because they can support nonzero stresses, which allow the two fluids to coexist at different pressures, and because mass transfer between phases occurs across these interfaces. The existence of multiple fluid phases in the pore space significantly complicates the physics and chemistry of the system, and the associated mathematical description becomes concomitantly more complex. These important pore-scale processes will ultimately be represented through parameters and variables defined on length scales much larger than the pore scale. Such larger scale equations are necessary for practical applications. In this chapter, we provide a description of the underlying porous medium system, highlighting the dominant processes and developing the mathematical expressions that will allow us to model many aspects of the CO_2 injection problem.

3.1 PHASES AND COMPONENTS

As described above, a distinguishing characteristic of two-fluid porous media systems is the existence of fluid–fluid interfaces at the pore scale. Figure 3.1 shows a schematic of a sample of porous medium in which two fluids exist. In these systems, the solid material tends to have a stronger surface attraction for one of the

Geological Storage of CO_2: Modeling Approaches for Large-Scale Simulation, First Edition.
Jan M. Nordbotten and Michael A. Celia.

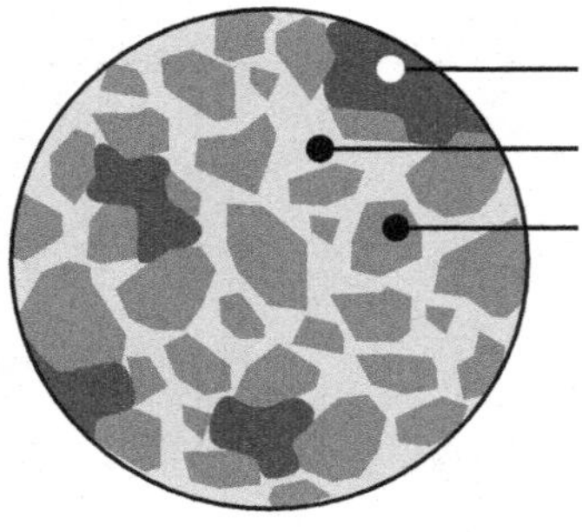

Figure 3.1 Fluid and solid distributions within an REV.

fluids. That fluid that is preferentially attracted by the solid is called the *wetting fluid*, while the other fluid is referred to as the *nonwetting fluid*. When one of the fluids is water, a solid that preferentially attracts water is called *hydrophilic*, while one that more strongly attracts the nonaqueous phase is called *hydrophobic*. At the location where the fluid–fluid interfaces meet the surface of the solid material, the angle formed between the interface and the solid is called the *contact angle*. The fluid on the side of the interface with an angle less than 90° with respect to the solid surface is the wetting fluid. Strongly wetting fluids tend to form contact angles close to zero.

Given the added complexity in this system, we need to expand the definitions introduced in Chapter 2. For example, while for single-fluid systems the porosity is identical to the fraction of the total volume occupied by the fluid, we now need to refine this measure to indicate the fraction of the pore space occupied by each of the fluids. To this end, we define the *fluid phase saturation*, s_α, which is defined as the fraction of the pore space occupied by fluid phase α, where α denotes either the wetting phase ($\alpha = w$) or the nonwetting phase ($\alpha = n$). The fraction of the total volume occupied by each phase is given by ϕs_α. Clearly s_α is bounded below by 0 and above by 1, $0 \le s_\alpha \le 1$, and if only two fluid phases are present, the sum of the fractions is also 1, $s_n + s_w = 1$.

In addition to the phase saturations, each phase has an associated volumetric flux vector, defined as the volume of phase α per total area of porous medium per time. The volumetric flux vector for phase α is denoted by $\boldsymbol{u}_\alpha$. Similar to the single-fluid-phase case, the actual fluid velocity is the volume of fluid per area occupied by phase α per time, so that $\boldsymbol{v}_\alpha = \boldsymbol{u}_\alpha / \phi s_\alpha$. Each of the fluid phases will have its own pressure, which we denote by p_α. Because the fluid–fluid interfaces can support nonzero stresses, a different pressure can exist on either side of the interface, and therefore, each phase will usually have a different pressure. We define the difference between the phase pressures as the *capillary pressure*, denoted by p^{cap}, with

$$p^{cap} \equiv p_n - p_w.$$

Before continuing, we note that all of these quantities are assumed to be defined at the laboratory (representative elementary volume [REV]) scale ℓ_{lab}. This means, in particular, that the fluid pressures are defined as a pressure over a volume with characteristic length scale ℓ_{lab}, and the capillary pressure is defined as the difference between these two laboratory-scale pressures. While recent research has addressed the suitability of this definition of capillary pressure, and other related issues associated with length scales and the definition of average pressure, we will simply take this standard definition of laboratory-scale capillary pressure for the remainder of this book.

Multi-fluid-phase systems are ubiquitous in the field of hydrology, soil science, and petroleum reservoir engineering. A simple example is the unsaturated soil zone, where the two fluids are air and water. In groundwater contamination problems, the contaminants often involve nonaqueous fluid phases such as organic solvents and hydrocarbon products. Such fluids are often referred to as non-aqueous-phase liquids (NAPLs). In petroleum reservoirs, oil and water are typically present, sometimes with natural gas as a third fluid phase. And, of course, if we inject CO_2 into a deep formation filled with brine, the CO_2 will exist as a separate fluid phase, thereby creating a multi-fluid-phase flow problem. For all of these problems, the mathematical tools discussed in Chapter 2 must be extended to account for the additional complexities associated with multiple fluid phases.

In many problems, the composition of a particular phase turns out to be important. For example, in the case of a NAPL problem, the existence of a nonaqueous phase in the subsurface is usually not a significant problem per se. Rather, it is the fact that groundwater usually flows through the subsurface zone containing the NAPL and in so doing, the water phase becomes contaminated due to interphase mass transfer involving specific components of the NAPL phase. The problem therefore becomes one of water contamination because of mass transfer of one or more components across the fluid–fluid interfaces involving the NAPL and water, and the subsequent movement of the contaminated water. If the NAPL is gasoline, then typical components of concern include benzene, toluene, and xylene, often referred to as BTX. If the NAPL is an organic solvent, then organics like trichloroethylene (TCE) are of most concern. NAPLs that allow small but significant amounts of their mass to dissolve into the aqueous phase are referred to as "slightly miscible." For the CO_2 problem, the CO_2 is slightly miscible with the resident brine, meaning that relatively small (but significant) amounts of CO_2 can dissolve into the brine. In addition, small amounts of H_2O can evaporate into dry CO_2. All of this means that in addition to the movement and behavior of the overall fluid phases, we also need to consider the composition of the fluids and account for different components within fluid phases that can transfer between or among the phases present. These interphase mass transfers can also include the solid phase.

Our overall view of the subsurface has now evolved into a system with a porous solid matrix within which the pore space is filled by one or more fluid phases. Each fluid phase is composed of a number of components, and the solid phase may involve a number of different minerals. These fluid phases move in the subsurface, carrying with them their respective components; components may change phases; and various reactions may take place involving some number of components and perhaps minerals.

We will assume a phase is composed of some number of components, with each component denoted with index i. We will use the *mass fraction*, m_α^i, as a basic measure of phase composition, where mass fraction of component i in phase α is defined as the mass of component i in phase α divided by the total mass of phase α. It is useful to be aware of the related measure of *mole fraction*, x_α^i, defined as the ratio of moles (or molecules) of component i in phase α to the total number of moles in phase α. The mass and mole fractions are related to one another through the molecular weights of the components, such that

$$m_\alpha^i = \frac{x_\alpha^i (MW)^i}{\sum_i x_\alpha^i (MW)^i},$$

where $(MW)^i$ denotes the molecular weight of component i. In a fairly simple model of CO_2 injection, we will have a three-component, two-phase system with components $i = \{CO_2, H_2O, \text{salt}\}$ (in reality a simplified representation for a number of components) and phases called brine ($\alpha = b$) and CO_2 ($\alpha = c$). While we will focus on the phases and components of direct relevance to the CO_2 problem, much of the background material in this chapter will apply to more general systems.

3.2 EQUATIONS OF STATE AND MASS TRANSFER

Equations of state usually relate density and viscosity of a particular phase to pressure, temperature, and composition, while mass transfer equations describe how component masses partition among multiple phases. While these equations can become quite involved, especially when considering general multicomponent, multiphase systems, in this section we focus only on the equations of state and equations of mass transfer for the CO_2–brine system.

Almost all CO_2 injection strategies to date involve injection into formations that are deep enough to have both temperature and pressure that exceed the critical point for CO_2, which is at approximately 31°C and 7.4×10^6 Pa or 7.4 megapascal (MPa). This means that CO_2 will be in a supercritical state, which means the CO_2 exhibits both gas-like and liquid-like properties—for example, it moves through small spaces like a gas but can dissolve materials like a liquid. Given a typical geothermal gradient of about 30°C/km and a surface temperature of about 25°C, the critical point is reached at a depth of about 800 m. Therefore, most injections are expected to take place below this depth. The phase diagram for CO_2 is shown in Figure 3.2.

In general, the properties of the fluids will depend on pressure, temperature, and composition. For CO_2, the main dependence is on temperature and pressure, while for brine, the composition will usually play an important role. Relationships for CO_2 density and viscosity as a function of pressure and temperature are shown in Figure 3.3. Notice that the properties have very strong variations around the critical point, but are then quite smooth and well behaved beyond the critical point. The specific

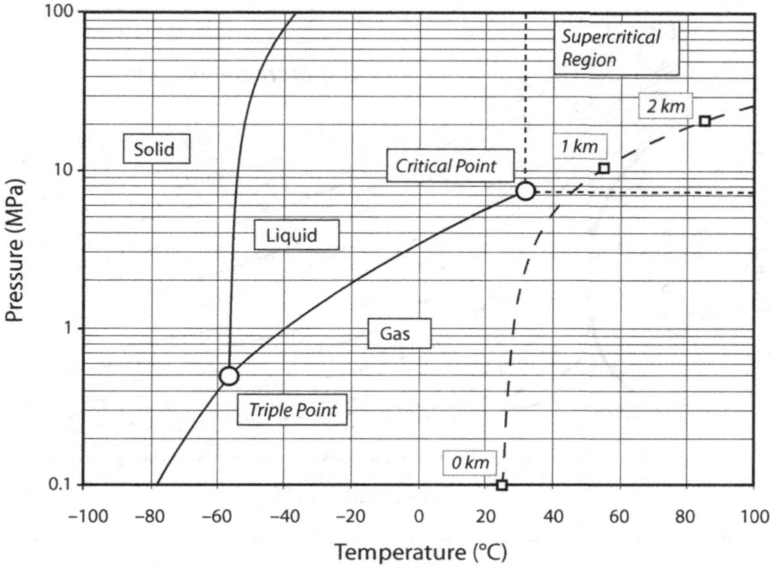

Figure 3.2 Phase diagram for pure carbon dioxide. The dashed line indicates typical temperatures and pressures as a function of depth below land surface.

functional forms that describe the relationships shown in Figure 3.3 are usually referred to as *equations of state*.

To give a sense of the values for density and viscosity over a range of subsurface conditions, we can consider "shallow" formations of 1-km depth and "deep" formation of 3-km depth, coupled with a range of geothermal gradients going from "cool" with a surface temperature of 10°C and geothermal gradient of 25°C/km to "warm" with a surface temperature of 20°C and geothermal gradient of 45°C/km. Tables 3.1 and 3.2 give the corresponding values of density and viscosity for both CO_2 and brine. The brine properties vary weakly with pressure, more strongly with temperature, and most strongly with composition (salinity). They are given for the assumed geothermal gradients coupled with a pressure gradient of 10.5 MPa/km and a range of salinities from 0.02 to 0.3, measured in mole fractions.

We note that the CO_2 density is always less than the brine density, by a difference $\Delta\rho = \rho_b - \rho_c$ ranging from 262 kg/m^3 to 944 kg/m^3. Viscosity contrasts, expressed as the ratio μ_c/μ_b, range from 0.026 to 0.2. So in all cases of CO_2 injection into brine-filled formations, the CO_2 will be much less dense and much less viscous than the resident brine.

As discussed in Section 3.1, components may change phase through interphase mass transfer. This phase change can have significant implications for problems like groundwater contamination. It can also have some potentially important effects in the CO_2 problem because CO_2 is slightly miscible with brine, and H_2O is likewise slightly miscible with CO_2. If we choose to model the dissolved salt in the brine,

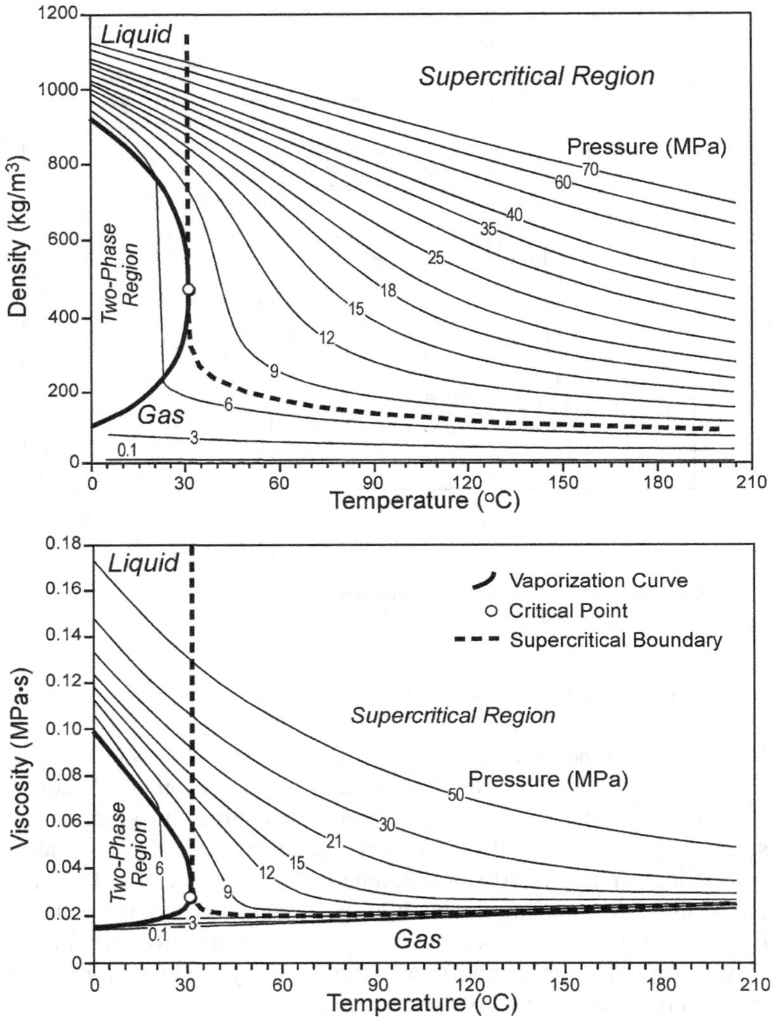

Figure 3.3 Dependence of CO_2 density and viscosity on temperature and pressure (from Nordbotten et al., 2005).

Table 3.1 Density Values for CO_2 and Brine (From Nordbotten et al., 2005)

	Cold basin (kg/m^3)	Warm basin (kg/m^3)
Shallow formation	$\rho_c = 714$ $\rho_w = 1012–1230$	$\rho_c = 266$ $\rho_w = 998–1210$
Deep formation	$\rho_c = 733$ $\rho_w = 995–1202$	$\rho_c = 479$ $\rho_w = 945–1145$

Table 3.2 Viscosity Values for CO_2 and Brine (From Nordbotten et al., 2005)

	Cold basin (MPa·s)	Warm basin (MPa·s)
Shallow formation	$\mu_c = 0.0577$	$\mu_c = 0.023$
	$\mu_w = 0.795–1.58$	$\mu_w = 0.491–0.883$
Deep formation	$\mu_c = 0.0611$	$\mu_c = 0.0395$
	$\mu_w = 0.378–0.644$	$\mu_w = 0.195–0.312$

we assume it is only miscible in the aqueous phase, and if the aqueous phase disappears (all H_2O evaporates into the CO_2 phase, for example), then the salt will precipitate as a solid phase. Dissolution of CO_2 into the brine phase has two consequences that may be important. The first is that it tends to lower the pH of the brine, due to the formation of carbonic acid. This lower pH can drive a sequence of geochemical reactions in the formation. The second is that it tends to increase slightly the density of the brine. This density difference can drive long-term convective mixing. The evaporation of H_2O into the CO_2 phase creates what is commonly called "wet CO_2", which is much more corrosive to metals than is dry CO_2. And the precipitation of solid salt into the pore space will reduce the porosity and concomitantly reduce the permeability of the medium. Any or all of these effects may be important, and therefore, we need to have tools to model interphase mass transfer for these systems.

While kinetic mass transfer expressions are sometimes required for groundwater contamination problems, for most other systems including deep hydrocarbon systems, equilibrium is usually assumed when considering phase partitioning. That is, given a mixture of components, each with a specified amount of mass, phases of specific composition are created via equilibrium partitioning relationships for the given conditions of pressure and temperature. For complex systems this is generally done through an imposed equality of chemical potential or phase fugacities. Because the CO_2–brine system is relatively simple, involving dilute solutions where one component dominates each phase, much simpler approaches may be used to capture the essential features of the interphase mass transfer. For example, in a simple CO_2–water system, the mass fraction of component CO_2 in the aqueous phase has been quantified as a function of pressure, temperature, and salinity. If the typical pressure and geothermal gradients used in Figure 3.2 are used to calculate pressure and temperature as a function of depth, we can estimate values of equilibrium mass fraction for CO_2 in the aqueous phase. For pure water, the value of $m_b^{CO_2}$ is about 0.047 at 1-km depth, increasing to about 0.06 at 3-km depth. Increasing salt content in the brine reduces the values of $m_b^{CO_2}$, a phenomenon known as the "salting out effect." A salt concentration of 5% (50,000 ppm) reduces these numbers by about 25%, ranging between 0.038 and 0.045, while a salt concentration of 30% leads to reductions of about 80%, so that the resulting values of $m_b^{CO_2}$ are all around 0.01. We find even smaller mass fractions for equilibrium partitioning of H_2O into a pure CO_2 phase, typically less than 1%.

3.3 TWO-PHASE EXTENSION OF DARCY'S LAW

Similar to the single-phase case, Darcy's law plays a central role in the mathematical description of multiphase flow. In Chapter 2, the simple experiments of Darcy were combined with concepts from fluid mechanics to provide the mathematical description of fluid movement. What began as a simple algebraic relation was extended to a differential expression, and then to anisotropic and heterogeneous materials. We now add another extension by considering flow in multi-fluid porous media systems. The standard extension of the Darcy equation to multiple fluids is based in large part on the schematic shown in Figure 3.1. The original version of Darcy's law applies to the system shown in Figure 3.1, under the condition that only one fluid (e.g., the wetting fluid) occupies all of the pore space. The hydraulic conductivity may then be seen as a measure of the energy loss associated with fluid flow through the tortuous pore spaces. In that case, all of the pore space is available for fluid flow. Now consider the case shown in Figure 3.1, where some of the pore space is occupied by one fluid, and the remainder of the pore space is occupied by a second fluid. From the point of view of the first fluid, the presence of the second fluid blocks some of the pore space, leaving the first fluid with a reduced set of pores through which to flow. This makes it more difficult for the fluid to flow, which is reflected in a lower value of the apparent permeability. To account for the reduction in permeability as a function of the presence of another fluid, a new factor is introduced into Darcy's equation to reduce the apparent permeability as a function of volumetric occupancy of the fluids—that is, as a function of the saturation s_α. This function is called the relative permeability function, and is denoted by $\boldsymbol{k}_{r,\alpha} = \boldsymbol{k}_{r,\alpha}(s_\alpha)$. Like the permeability, the relative permeability will in general be anisotropic. With this new function included, Darcy's law for multiphase flow is now written as a direct extension of Equation (2.10), which we write as follows:

$$\boldsymbol{u}_\alpha = -\frac{\boldsymbol{k}_{r,\alpha}\boldsymbol{k}}{\mu_\alpha}(\nabla p_\alpha + \rho_\alpha g \mathbf{e}_z) = -\frac{\boldsymbol{k}_{r,\alpha}\boldsymbol{k}}{\mu_\alpha}(\nabla p_\alpha - \rho_\alpha \boldsymbol{g}). \tag{3.1}$$

This extension of Darcy's equation uses a single function, the relative permeability function, to represent a number of complex interactions at the pore scale, in particular involving fluid–fluid interfaces. The construction of Equation (3.1) is based on the view that fluids blocking pore spaces can be treated as if they were solids. That is, a given configuration of fluids at the pore scale is taken to represent a particular state of the system, and for that small-scale geometric configuration, an effective conductance for the fluid phase of interest can be calculated, assuming everything else in the system is static. The ratio of that conductance to the conductance at full saturation gives the value of relative permeability at that saturation. Of course, fluid–fluid interfaces can be quite different than fluid–solid interfaces, and the pore-scale flow processes can become quite complex. The multiphase flow system is significantly different than Darcy's original experimental system, so this multiphase extension of Darcy's law may be viewed with some skepticism, and indeed this is an area of active research. However, the general behavior of the system is often represented well by Equation (3.1). While acknowledging ongoing discussion in the research community about the appropriate form of flux laws for multi-fluid porous media systems, we will use Equation (3.1) as a practical modeling tool, especially as applied to the CO_2 problem, where the dominant physical processes in the system make Equation (3.1) a reasonable model to describe fluid fluxes.

It is clear both conceptually and through experimental evidence that the relative permeability function is anisotropic. This is most easily understood by considering the subscale layering (or heterogeneity) that gives rise to anisotropic permeability (recall Box 2.1). These subscale patterns, in the presence of capillary forces, will likely also correlate with subscale patterns of fluid distribution, which allows the fluids to preferentially flow in one or the other direction. Nevertheless, relative permeability is almost universally approximated as an isotropic (scalar), nonlinear function of saturation. This leads to simplifications both in the manipulation of equations as well as in numerical methods. In this chapter, we will use boldface to emphasize the physics of anisotropic relative permeability when this does not lead to complications, and use the standard simplification to scalar relative permeability whenever necessary. A typical form of the isotropic relative permeability function is illustrated in Figure 3.4, where several important features are emphasized. The first is that the curves in the figure are nonlinear, so that relative permeability is a nonlinear function of saturation. The second is that the relative permeability curves for each of the phases go to zero well before the actual phase saturation goes to zero. The saturation at which the relative permeability goes to zero is referred to as the *residual saturation*, which we denote by s_α^{res}. Again referring to the pore-scale description illustrated in Figure 3.1, the residual saturation state usually implies that some amount of the phase is present within the averaging volume but the phase fails to form a continuous pathway through which flow of that phase can take place. That is, the remaining fluid tends to be in isolated regions whose size is small relative to the averaging volume.

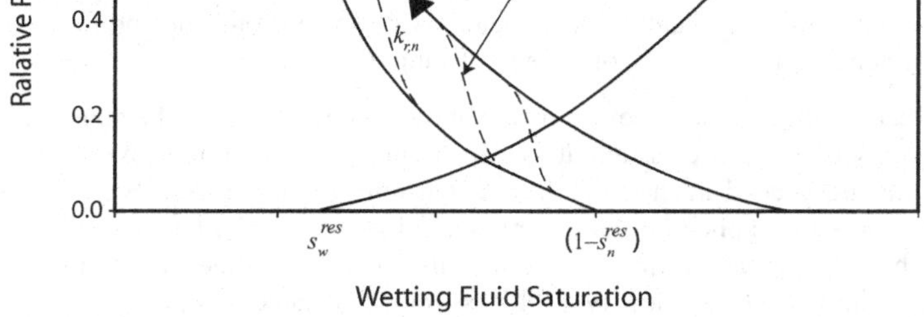

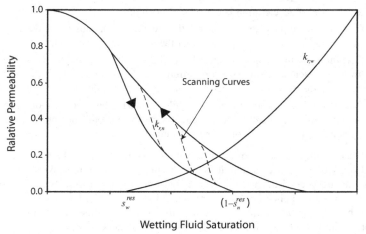

Figure 3.4 Typical forms for relative permeability curves.

Note that it is this concept that gave rise to the notion of drainable porosity in Chapter 2. Finally, we note that, as drawn in Figure 3.4, the relative permeability curve for the nonwetting phase exhibits hysteresis—that is, a different curve is followed depending on whether the saturation is increasing or decreasing. The appropriate directions are denoted by the arrows superimposed on the curves. When the rate of change of saturation changes sign away from the end points of the relative permeability curve, models are needed for the transition between the two curves. We will see this again in the capillary pressure curves of the next section.

To complete this introduction to the extended version of Darcy's law, we introduce one more variable, the *phase mobility*, $\boldsymbol{\lambda}_\alpha$, which is defined as the ratio of the relative permeability function to the phase viscosity, that is, $\lambda_\alpha(s_\alpha) \equiv (k_{r,\alpha}(s_\alpha))/\mu_\alpha$, where the dependence on saturation is shown for both the mobility and the relative permeability. With this definition Darcy's law may be written as follows:

$$\boldsymbol{u}_\alpha = -\lambda_\alpha \boldsymbol{k}(\nabla p_\alpha - \rho_\alpha \boldsymbol{g}). \tag{3.2}$$

We will use both Equations (3.1) and (3.2) to represent Darcy's law for multiphase flow systems (see Box 3.1).

BOX 3.1 *Two-Phase Darcy's Law*

The natural way to think of Darcy's law is as a balance of forces. With this objective, we rearrange Equation (3.2) as

$$(\lambda_\alpha(s_\alpha)\boldsymbol{k})^{-1}\boldsymbol{u}_\alpha = -\nabla(p_\alpha + \rho_\alpha g z).$$

The left-hand term is interpreted as friction forces due to flow, while the right-hand side represents the force due to the gradient of the fluid potential. This balance of forces is based on several assumptions:

1. Balance of forces implies that the fluids move at a terminal velocity, thus the time scale of acceleration is considered negligible.
2. The flux $\boldsymbol{u}_\alpha$ represents motion relative to the solid, thus we have implicitly chosen a frame of reference such that the solid phase is stationary.
3. The momentum transfer associated with the fluid–fluid interfaces would be naturally represented by a term proportional to the relative velocity $\boldsymbol{u}_\alpha - \boldsymbol{u}_\beta$. By omitting this term these forces are assumed to be negligible.
4. The saturation dependent mobility, while allowed to be dependent on the history of saturation, is usually not modeled as dependent on other properties of the system, such as the rate of change of saturation.

Through careful experiments, porous media and flow conditions can be constructed that invalidate any of the above assumptions. This implies that conditions exist where additional terms are needed in the two-phase extension of Darcy's law. Nevertheless, for practical large-scale applications, all four assumptions are usually taken to be valid. This will also be our approach in this book, since the large time scales and fluid properties associated with CO_2 storage justify all the assumptions stated above.

3.4 CAPILLARY PRESSURE AND PORE-SCALE MODELS

As was the case in the discussion of single-phase flow in Chapter 2, Darcy's law by itself is not sufficient to describe the two-phase porous media flow system. We need to add additional equations and constraints, which take the form of mass balance equations, a relationship between the phase pressures and the phase saturations, and a simple relationship involving the phase saturations.

3.4.1 Capillary Pressure

Capillary pressure is defined as the difference in phase pressures, and is most commonly parameterized as an algebraic function of phase saturation. This relationship is taken to be algebraic because it is often based on experiments that take measurements only when interfaces are static and equilibrium conditions have been reached, or on analyses of dynamic data that impose the assumption that the relationship between saturation and capillary pressure is algebraic. It is instructive to consider briefly a typical measurement of this relationship in a static cell called a pressure cell. Assume we have a sample of porous medium whose pore space is initially completely filled with wetting fluid. Assume the lateral sides of the sample are impermeable. The bottom boundary is connected to a reservoir of wetting fluid, whose pressure can be controlled, and the top boundary is connected to a nonwetting fluid reservoir whose pressure can also be controlled. Assume the pressure cell is small enough that the difference in pressure between the top and bottom of the column (due to gravity) can be neglected. Finally, assume that the initial pressure in each of the fluid reservoirs is the same. The initial state is that all pores are filled with wetting fluid, and along the top surface there are fluid–fluid interfaces that are stable.

When these interfaces experience a pressure difference between the fluids on either side of the interface, the interface deforms and achieves a curvature that is given by the Young–Laplace equation, which is based on a balance of forces. If we represent a given pore as a cylindrical tube with radius r_{eff}, then the maximum capillary pressure that can be supported within a tube of that radius is given by the following equation:

$$(p_n - p_w)_{max} = (p^{cap})_{max} = \frac{2\gamma\cos\theta}{r_{eff}}, \tag{3.3}$$

where γ is the interfacial tension between the two fluids and θ is the contact angle that the interface makes with the solid surface of the pore. For capillary pressures below this maximum value, the interface has less deformation (less curvature), and when there is no pressure difference, the interface will be flat. Note that a porous medium will have a range of pore sizes, giving rise to a distribution of values for the radii (or effective radii) r_{eff}. This means that different pores will have different values of $(p^{cap})_{max}$. Note that for CO_2–brine systems, measurements indicate that the interfacial tension varies between 25 milli-Newton per meter (mN/m) and 45 mN/m, depending on the pressure, temperature, and brine composition. For comparison, air–water interfacial tension is about 70 mN/m, which is a relatively high value of interfacial tension.

Now consider an experiment where the nonwetting phase pressure is increased by some increment, call it Δp_n, while the wetting phase pressure is held constant, and the system is allowed to equilibrate. The increase in nonwetting phase pressure results in a corresponding increase in capillary pressure. If this initial increase is smaller than the maximum values for all of the interfaces, then the only response will be that all of the interfaces at the top of the sample deform slightly to achieve a curvature in proportion to the pressure increase. At some point in this progression of pressure increases, one or more of the interfaces will exceed its maximum capillary pressure, and the interface in that pore will become unstable and displace into the sample. It will continue to advance, perhaps splitting into multiple new interfaces, until all of the interfaces wind up in pores that are small enough to support the imposed capillary pressure. The displacement will cause wetting fluid to leave the sample through the bottom boundary. Measurement of the amount of fluid that leaves gives a measure of the change in saturation within sample. Once outflow ceases, equilibrium is assumed to exist, and a data point relating the capillary pressure to the saturation is recorded. This is then continued, stepwise, until a state is reached where further increases in applied pressure do not result in any change in saturation. The experiment can then be repeated by successively decreasing the nonwetting fluid pressure and recording the amount of wetting fluid that reenters the sample. The data pairs of capillary pressure and saturation can then be plotted, with a typical (idealized) result shown in Figure 3.5.

Referring to the curves in Figure 3.5, we make the following four points. First, when nonwetting fluid displaces wetting fluid, the process is referred to as *drainage*. When wetting fluid displaces nonwetting, it is called *imbibition*. Second, as with the relative permeability function, we again see clearly the residual saturation values for both the wetting and nonwetting phases. The values of these residuals are important, especially in the CO_2–brine system, because residual brine saturation can make injection of CO_2 more difficult, and residual CO_2 saturation is an important trapping mechanism. The third point is that the relationship between capillary pressure and saturation tends to be strongly hysteretic, with obvious differences between drainage and imbibition curves. A curve that includes full saturation as one of its end points is called a *primary* displacement curve—the *primary drainage* curve is shown in the figure. Primary drainage is also the initial curve described in the previous paragraph. Curves that begin at the residual saturation point of the other fluid are referred to as *main* displacement curves. Figure 3.5 shows both *main drainage* and *main imbibition* curves. Furthermore, curves that begin at points between the two residual saturations are referred to as *scanning curves*. The fourth point is that the primary and main drainage curves show that a finite capillary pressure is required before any drainage displacement begins. This is consistent with the description given in the previous paragraph, where a finite imposed pressure was required to displace the

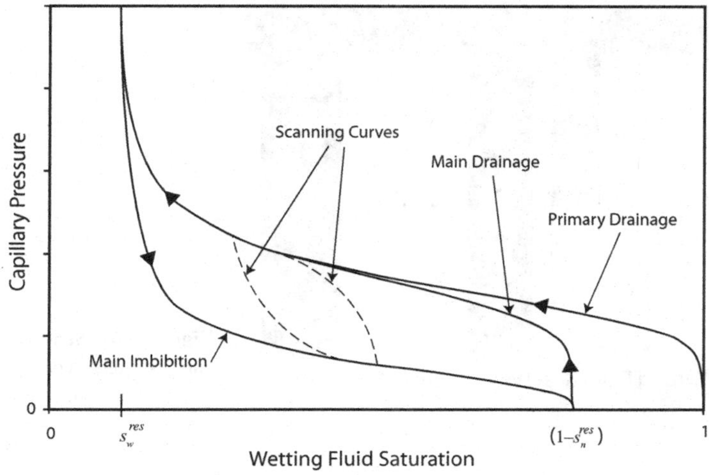

Figure 3.5 Typical form for a capillary pressure–saturation curve.

first interface in the pore space. The capillary pressure required to initiate displacement of the wetting fluid by the nonwetting fluid is called the *entry pressure*. The phenomenon whereby nonwetting fluid is unable to enter particular spatial regions that are filled with wetting fluid because of failure to reach this entry pressure is referred to as *capillary exclusion*. Capillary entry pressure and capillary exclusion can be important phenomena in the carbon storage problem.

3.4.2 Pore-Scale Models

The procedure of sequential pressure increases that drive interface motion can be quantified through the use of pore-scale models that explicitly track fluid–fluid interfaces through modeled pore spaces. The simplest of these is referred to as a "bundle-of-tubes" model. In this model, the pore space is represented by a set of parallel and nonintersecting capillary tubes, each having its own radius (see Figure 3.6). Assume that one end of the tubes is connected to a reservoir of wetting fluid while the other end is connected to a reservoir of nonwetting fluid, and that fluid pressure can be controlled in each of the reservoirs. If the reservoirs are initially set to the same pressure, and if the system is sufficiently small that gravity effects are negligible, then the initial phase pressures are equal and the capillary pressure is zero. Under this condition, the tubes will all be filled with wetting fluid, so the initial wetting phase saturation is equal to one.

Consider a computational experiment in which the wetting fluid reservoir is held at constant pressure and the nonwetting fluid pressure is increased incrementally. In order to describe the response of the system, let us number the capillary tubes in order of descending radius, such that tube 1 has the largest radius, tube 2 the second-

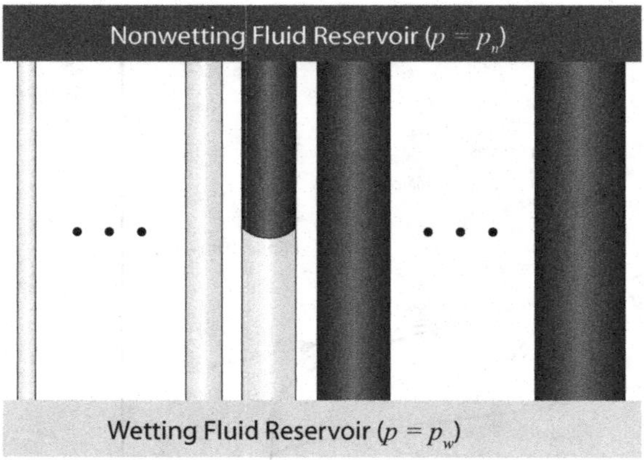

Figure 3.6 Schematic of a bundle-of-tubes conceptualization of a porous media.

largest, and so on, through to tube N which has the smallest radius. As the nonwetting fluid pressure is increased, no tubes drain until the entry pressure for the largest capillary is exceeded. From Equation (3.3), we know this occurs at a capillary pressure of

$$(p^{cap})_{max,1} = \frac{2\gamma\cos\theta}{r_{eff,1}},$$

where subscript 1 denotes tube number 1. If areal fractions are used as a measure of phase saturation, then the wetting phase saturation after the first tube has filled is given by

$$s_w\left((p^{cap})_{max,1}\right) = 1 - \frac{\pi r_{eff,1}^2}{\sum_{i=1}^{N} \pi r_{eff,i}^2}.$$

The procedure continues systematically, just like the physical experiment we are mimicking. Plotting the pairs of capillary pressure–saturation data points results in a capillary pressure–saturation curve whose shape reflects the pore-size distribution. A typical result from this kind of simple bundle-of-tubes model is shown in Figure 3.7. Note that in the limit of a large number of tubes, the discrete calculations can be replaced by continuous functions and the resulting capillary pressure–saturation relationship is then smooth and continuous.

Whether derived using discrete calculations or continuous functions, the resulting relationship between capillary pressure and saturation has the following characteristics. The relationship exhibits finite entry pressure, but there are no residual saturations for either phase, and there is no hysteresis (unless contact-angle hysteresis is imposed—i.e., the contact angle is different depending on whether the displacement is drainage or imbibition). The reason for the lack of both hysteresis and residual saturations is the overly simplistic geometries used in the model. In particular, the lack of intersections, which are present in the interconnected, tortuous

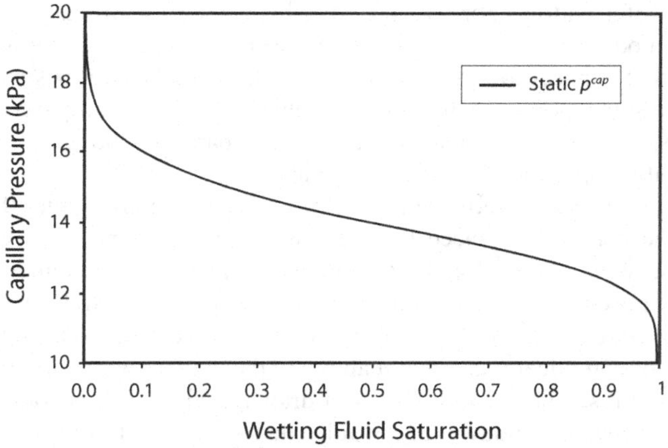

Figure 3.7 Capillary pressure curve resulting from a bundle-of-tubes model.

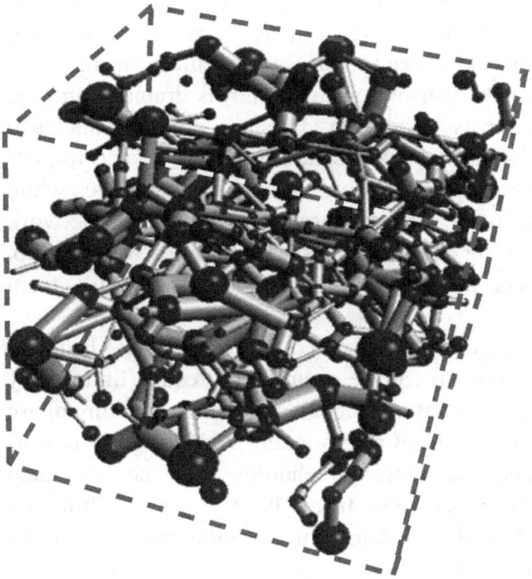

Figure 3.8 Conceptualization of a network model based on the digital rock shown in Figure 2.2 (courtesy of Numerical Rocks A/S).

pathways that characterize real porous media, lead to these overly restrictive results.

One way to improve on the bundle-of-tubes model is to create interconnected pore structures that more closely represent actual porous media. This is often done using so-called pore network models. Instead of parallel and nonintersecting tubes, network models create a network of interconnected tubes like the network shown in Figure 3.8. The individual tubes are often referred to as pore throats, and the junctions are referred to as pore bodies. Pore bodies are usually larger pore spaces, and in these models a pore body is typically restricted to be larger in size than the pore throats that connect to it. As opposed to simple bundle-of-tubes models which are based on a single pore-size distribution, network models usually have a pore-size distribution specified for the pore throats and a separate, although possibly correlated, distribution specified for the pore bodies.

Within a pore network model, the sequence of calculations mimics those already described. The network is placed between two reservoirs, one filled with wetting fluid and one with nonwetting fluid, and the sequence of incremental pressure changes is imposed. Each fluid–fluid interface is tracked through the network—a process that is roughly analogous to the bundle-of-tubes calculations—but now there are a number of complications brought about by the network structure. For example, some regions of resident (wetting phase for drainage) fluid may become completely surrounded by invading (nonwetting) fluid. The surrounded fluid becomes trapped, or isolated, because it no longer has any connected pathway to its home reservoir and therefore cannot drain out of the network. This trapping leads to residual saturations. Therefore, network models produce nonzero residual saturations.

Because of the restriction that each pore body must be larger than the pore throats that connect to it, pore bodies will never restrict interface movements during drainage. Conversely, pore throats will not restrict interface movement during imbibition. As such, it is the pore throat size distribution that controls drainage and the pore body size distribution that controls imbibition. Because pore throats have smaller radii than pore bodies, and capillary pressures are inversely related to pore radius (Eq. 3.3), the drainage curve corresponds to higher capillary pressures while the imbibition curve corresponds to lower capillary pressures. Therefore, network models produce hysteresis in the resulting capillary pressure–saturation relationship, independent of the existence of contact-angle hysteresis. The hysteresis results from the network structure itself.

A typical result from a network model is presented in Figure 3.9. All of the important characteristics of the relationship can be seen in the result (finite entry pressure, hysteresis, residual wetting phase saturation and residual nonwetting phase saturation). These kinds of results can represent a reasonably wide range of porous media, especially given recent advances in imaging technology and the associated analysis tools to extract realistic pore networks from those images. While our description has been of simple models with cylindrical throats, modern pore network models use angular shapes so that wedges of wetting fluid remain in the corners of the pore elements (throats and pores) after a drainage event, thereby providing some legacy connectivity for the wetting phase through small wedges and films.

In addition to the capillary pressure–saturation relationships, these pore-scale models can also be used to estimate the relative permeability function. The idea is based on the computational analog to Darcy's original experiment. Consider a pore network model where only one fluid (e.g., the wetting fluid) is present, and both reservoirs are also filled with the wetting fluid. Because the pore throats are smaller than the pore bodies, they will offer more resistance to flow. If we assume all of the resistance comes from the pore throats, and the flow in all pore throats is laminar,

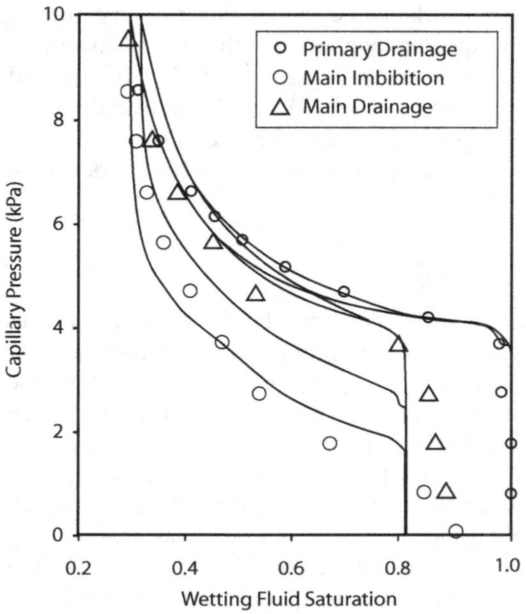

Figure 3.9 Capillary pressure curve from a network model, as compared to experimental data. Figure modified from Fischer and Celia (1999).

then the flow in the tube representing the pore throat can be assumed to have a parabolic velocity profile consistent with Pouiselle flow. The volumetric flow through the tube is linearly proportional to the difference in pressures at the ends of the tube, which corresponds to the pressures in the pore bodies at the two ends of the tube. That is, if a pore throat is connected to pore bodies i and j, and has radius $r_{eff,i,j}$ and length $l_{i,j}$, then the volumetric flow rate through the tube is given by

$$q_{i,j} = \frac{\pi r_{eff,i,j}^4}{8\mu} \frac{p_i - p_j}{l_{i,j}}.$$

Conservation of mass can be enforced at each pore body through the constraint that the inflows and outflows must balance, that is, for each pore body i the flow rates must satisfy

$$\sum_j q_{i,j} = 0,$$

where the sum is taken over all pore bodies j that are connected by a throat to pore body i, and constant fluid density is assumed. If this mass balance equation is written for each pore body, and the expression for the flows in terms of pressures at the pore bodies is substituted along with the boundary conditions associated with the two reservoir pressures and the lateral no-flow boundaries, the result is a system of linear equations that can be solved for the pressures at each pore body. Therefore, all of the characteristics of the flow field within the network can be determined.

These resulting flows within the network can be summed over a cross-section of the network and used in the Darcy equation, applied over the entire network, to estimate the permeability or hydraulic conductivity for the network. Let the summed flows over the cross section of the network be denoted by Q. If the overall length between the two reservoirs is L, the pressure difference between the reservoirs is ΔP, and the cross-sectional area is A, then the permeability for the network can be calculated from Darcy's law by

$$k = -\frac{Q\mu L}{A\Delta P}.$$

The same idea can be applied to calculate relative permeability values. The procedure takes advantage of the fact that at any given step in the capillary displacement calculations, the location of all fluid–fluid interfaces is known, and therefore, the fluid occupancy of each pore element (body and throat) is also known. For a given pattern of fluid occupancy, the network can be (computationally) extracted from the capillary pressure experiment and placed in a Darcy experiment cell, where both of the reservoirs are filled with the same fluid, and the flow through the connected pathways of that fluid is simulated using the same procedure as that already described. The regions occupied by the other fluid are considered to be blocked to flow of the fluid in the reservoirs, and this needs to be reflected properly in the connected flow pathways. The result is reduced bulk conductance across the network, due to the reduction in flow pathways. When the permeability for the phase of interest is calculated, it can be compared (in a ratio) to the intrinsic permeability; that ratio is a direct measure of relative permeability. Notice that at residual saturation, the phase at residual will have zero permeability because it will have no connected pathways through which fluid can flow from one reservoir to another. An example of relative permeability curves calculated from pore network models is given in Figure 3.10. Note that the model used to produce these results included angular pore elements and a careful construction of the pore space based on high-resolution images.

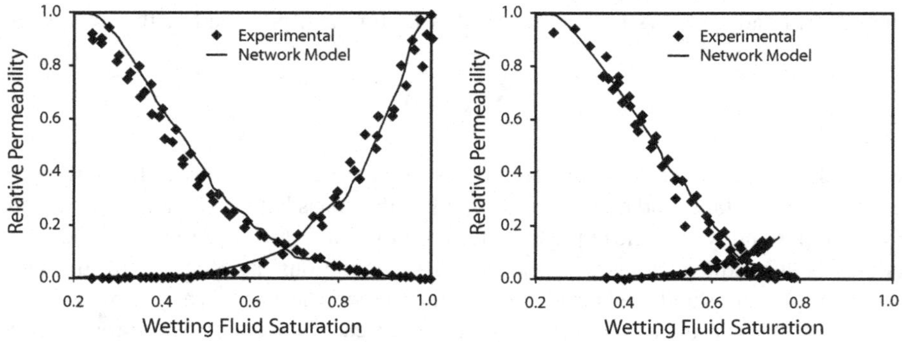

Figure 3.10 Relative permeability curves from a complex network model (adapted from Piri and Blunt, 2005).

All of the pore-scale models presented in this section have assumed a "quasi-static" displacement procedure. That is, if a fluid–fluid interface is determined to be unstable, it is moved through the network to a stable position instantaneously, and the capillary displacement algorithm includes no consideration of the time dimension. If the transient behavior of the displacement process is of interest, then so-called dynamic pore-scale models need to be developed. Such models can give rise to dynamic capillary pressure, and an associated dynamic equation to replace the static, algebraic relationship between capillary pressure and saturation. While this is an interesting research area, we will not consider it further. This is motivated by our analysis of time scales in the context of the CO_2 problem, which is presented in Chapter 4 and indicates that these small-scale dynamic effects are unlikely to be important for large-scale CO_2 injection and migration.

3.4.3 Capillary Pressure and Relative Permeability Relationships for CO_2–Brine Systems

The porous media system with brine and CO_2 as the two fluid phases is an example of a two-phase flow system. As such, data will be required to relate relative permeability and capillary pressure to saturation. These data should be expected to follow the kinds of curves we have already seen. One example of data measured specifically using brine and supercritical CO_2 is shown in Figure 3.11. These data are measured using rock samples from two different formations in the Alberta Basin: the Nisku Formation (a carbonate rock formation) and the Basal Sandstone Formation (a sandstone formation). The capillary pressure curves shown in the figures were measured under drainage conditions, and the saturation axis has been normalized using the effective saturation, which is defined for the brine as

$$s_{b,N} = \frac{s_b - s_b^{res}}{1 - s_b^{res} - s_c^{res}}.$$

A similar expression would be used to define the effective saturation of CO_2.

While the residual saturation values for brine are around 30% for both formations, the effect on relative permeability for the CO_2 is noticeably different. The Basal Sandstone has its permeability to CO_2 reduced by about 50% compared to the intrinsic permeability, while the Nisku has its permeability to the CO_2 reduced by about 80%. These reductions can have an important impact on the injection rate for CO_2, and directly affect the type of calculations we present in Chapters 5 and 6.

3.5 GOVERNING EQUATIONS AND SIMPLIFICATIONS

We complete the description of two-phase flow in porous media with mass conservation equations, analogous to Equation (2.15) in Chapter 2. While somewhat more complex because of the multiple phases and components involved, the overall

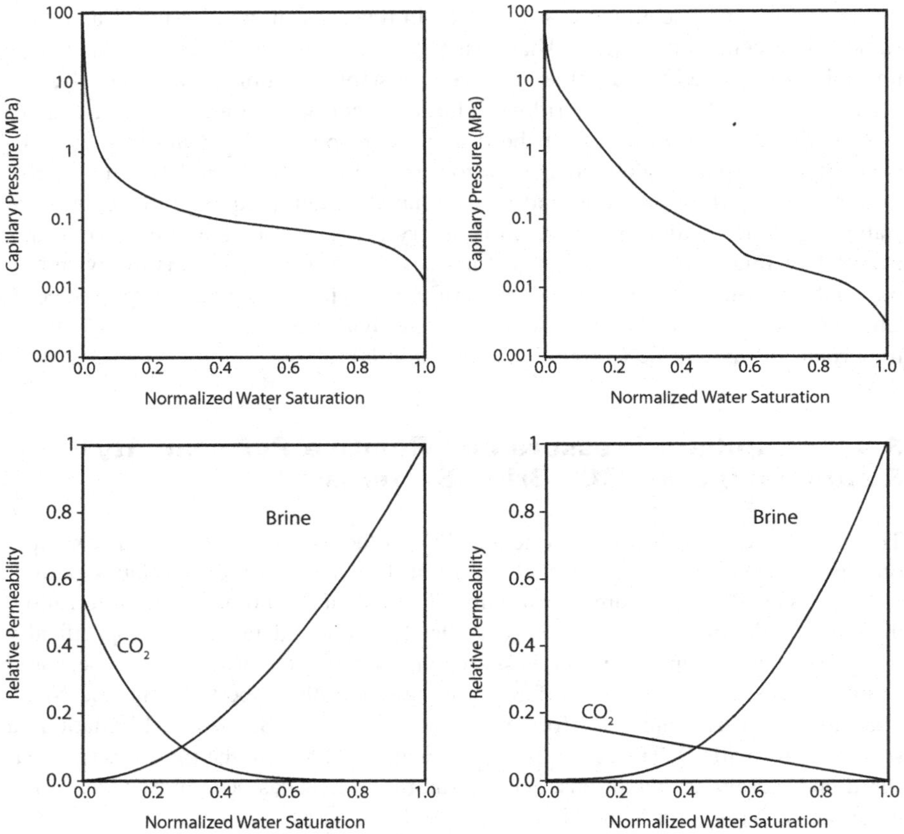

Figure 3.11 Capillary pressure and relative permeability curves for brine–CO_2 systems (modified from Bennion and Bachu, 2008).

approach is analogous to that used in Chapter 2. This includes the idea that these equations are based on (conserved and nonconserved) variables that are defined at the laboratory scale, ℓ_{lab}, and the conservation and balance equations are derived so that partial differential equations result. We also follow Chapter 2 in presenting some of the most common simplifications and linearizations of the governing equations.

3.5.1 Component Mass Conservation Equations

We write a general mass conservation equation for each component i in the system, following the general mass balance equation introduced in Chapter 2 (see Eqs. 2.12–2.16). To account for all of the mass of a component, we sum over all phases α, leading to the following terms in the general balance equation:

$$m^i = \sum_\alpha \rho_\alpha \phi s_\alpha m_\alpha^i, \quad f^i = \sum_\alpha \rho_\alpha \boldsymbol{u}_\alpha m_\alpha^i + \boldsymbol{j}_\alpha^i, \quad \text{and} \quad \psi^i = \sum_\alpha \psi_\alpha^i.$$

When substituted into the general conservation equation, we have the following equation for each component of the system:

$$\sum_\alpha \left[\frac{\partial(\rho_\alpha \phi s_\alpha m_\alpha^i)}{\partial t} + \nabla \cdot (\rho_\alpha \boldsymbol{u}_\alpha m_\alpha^i + \boldsymbol{j}_\alpha^i) \right] = \sum_\alpha \psi_\alpha^i \tag{3.4}$$

or

$$\frac{\partial m^i}{\partial t} + \nabla \cdot \boldsymbol{f}^i = \psi^i, \tag{3.5}$$

where m^i is the total mass (summed over all phases) of component i per volume of porous medium, $\boldsymbol{f}^i$ is the total mass flux vector for component i, and ψ^i represents the addition of mass of component i, defined as mass of component i added to the fluid phases per total volume per time. Equation (3.5) is a deceptively simple-looking equation; the behavior of components can be quite complex in terms of their distribution among phases as well as the equations representing the total mass flux vectors which include phase-based advective fluxes as well as non-advective fluxes.

As mentioned in Chapter 2, the non-advective flux term $\boldsymbol{j}_\alpha^i$ is often written using a form of Fick's law, where the flux is proportional to the gradient of the mass fraction, with the coefficient of proportionality being a diffusion (or diffusion-like) coefficient. When the process is dominated by sub-averaging-scale fluctuations in the velocity field (in this case velocity variations below the scale ℓ_{lab}), the coefficient tends to be much larger than the molecular diffusion coefficient and is usually referred to as the dispersion coefficient. On larger scales, it becomes known as macro-dispersion. Dispersion is usually an anisotropic process, so the dispersion coefficient actually turns out to be a matrix of values (a so-called tensor), analogous to the permeability matrix (tensor) in Darcy's equation.

Similar to the single-phase case, the mass balance equation, in this case Equation (3.4), does not constrain the system sufficiently to allow for solution of the unknown variables. Additional, material-specific equations must also be used. In the case of single-phase flow, these included Darcy's law and the equations for compressibility. For component transport in single-phase systems, this also included the ideas of dispersion and dispersion coefficients. To obtain additional equations for the multiphase case, we apply similar equations, although the existence of multiple phases makes the equations inherently more complex. In particular, we apply the multiphase version of Darcy's law, and we also apply the relationship between capillary pressure and saturation. In addition, because we are dealing with components, we need to determine how a particular component is distributed among the phases. To do this, we often apply equilibrium relations that specify the distribution of (at least some of) the components among the fluid phases. This is often referred to as a flash calculation. While these flash calculations are reasonable for local-scale problems, for large-scale CO_2 injection problems, equilibrium partitioning may not be appropriate, and we may need to consider some kinetic expressions.

If kinetics are needed, then we will usually need to write mass balance equations for specific components within a given phase. Therefore, we write a balance equation analogous to Equation (3.4), but without the sum over phases. The equation has the following typical form:

$$\frac{\partial(\rho_\alpha \phi s_\alpha m_\alpha^i)}{\partial t} + \nabla \cdot (\rho_\alpha \boldsymbol{u}_\alpha m_\alpha^i + \boldsymbol{j}_\alpha^i) = \psi_\alpha^i + \sum_{\beta \neq \alpha} \psi_{\alpha,\beta}^i. \tag{3.6}$$

Here, we have introduced $\psi_{\alpha,\beta}^i$ to represent kinetic mass exchanges of component i between phase α and all other phases, which are referred to by β. The term ψ_α^i represents external sources or sinks for component i. We note that for massless interfaces between fluid phases, the mass of component i moving from phase α to phase β must be equal and opposite the mass moving from phase β to phase α, in other words $\psi_{\alpha,\beta}^i = -\psi_{\beta,\alpha}^i$.

For systems with relatively simple component behavior, we often write mass balance equations for phases rather than components. This can be achieved by direct analysis of the phases or by summation of the component equations over all components in a phase. Beginning with the component phase balance equation, Equation (3.6), we can sum over all components to arrive at a mass balance equation for fluid phase α that has the following form:

$$\frac{\partial(\rho_\alpha \phi s_\alpha)}{\partial t} + \nabla \cdot \left(\rho_\alpha \boldsymbol{u}_\alpha + \sum_i \boldsymbol{j}_\alpha^i \right) = \psi_\alpha + \sum_{\beta \neq \alpha} \psi_{\alpha,\beta}. \tag{3.7}$$

Equation (3.7) takes on particular significance for immiscible fluids with no initial concentration gradients, since the dispersive flux and phase interaction terms are then zero. In this case, the mass in each phase is a conserved quantity, satisfying

$$\frac{\partial(\rho_\alpha \phi s_\alpha)}{\partial t} + \nabla \cdot (\rho_\alpha \boldsymbol{u}_\alpha) = \psi_\alpha. \tag{3.8}$$

We will use this conservation equation for immiscible flow of multiple fluids in porous media extensively in this and the following chapters.

3.5.2 Common Simplifications

There are many ways to think about simplifying the equations for multiphase, multicomponent transport to more manageable forms, both for analytical treatment and numerical simulation. We saw these ideas already in Section 2.2.2 for single-phase flow, where we introduced linearized equations with compressibility replacing more general equations involving an equation of state. Here we review some of the historically more popular, and important, modeling simplifications as applied to two-phase flow.

Phase Equations with Compressibility

When writing equations for phases, we use the balance equation given above coupled with Darcy's extended equation for multiphase systems, the relationship between capillary pressure and saturation, and compressibility arguments for the fluids and the solid matrix that are analogous to those presented in Chapter 2. Consider the time derivative term in the phase balance Equation (3.8). Because phase saturations and pressures are usually our independent variables of choice, we are motivated to relate the density and porosity to those variables, which in this case means the fluid pressures. We use arguments analogous to those used in Section 2.2.2 to write the time derivative as follows:

$$\frac{\partial(\rho_\alpha \phi s_\alpha)}{\partial t} = \rho_\alpha \phi \frac{\partial s_\alpha}{\partial t} + s_\alpha \frac{\partial}{\partial t}(\rho_\alpha \phi) = \rho_\alpha \phi \frac{\partial s_\alpha}{\partial t} + s_\alpha \rho_\alpha c_\alpha \frac{\partial p_\alpha}{\partial t}, \tag{3.9}$$

where c_α is the total compressibility (porosity and fluid) coefficient associated with phase α.

Note that if we combine Darcy's law (Eq. 3.2) with the mass balance equation for a phase (Eq. 3.8), and if we assume the densities, viscosities, and relative permeability functions are all known, we then have two equations (Eq. 3.8) for the two fluid phases, say $\alpha = w$ and $\alpha = n$. But we have four unknowns in the equations: two saturations and two phase pressures. These remain as the four unknowns whether or not we include compressibility effects, assuming we know the compressibility coefficients. We recall that a third equation is given by the simple geometric equality that says the pore space must always be completely filled with fluids, that is,

$$\sum_\alpha s_\alpha = s_w + s_n = 1. \tag{3.10}$$

We still need a fourth equation, which turns out to be the relationship between the saturation s_w and the capillary pressure $p^{cap}(s_w)$. Therefore, we can write the equations for a standard (and relatively simple) two-fluid system as Equation (3.10) combined with the following three equations:

$$\rho_\alpha \phi \frac{\partial s_\alpha}{\partial t} + s_\alpha \rho_\alpha c_\alpha \frac{\partial p_\alpha}{\partial t} - \nabla \cdot [\rho_\alpha \boldsymbol{\lambda}_\alpha \boldsymbol{k}(\nabla p_\alpha - \rho_\alpha \boldsymbol{g})] = \psi_\alpha \quad (\alpha = w, n). \tag{3.11}$$

$$p_n - p_w = p^{cap}(s_w). \tag{3.12}$$

These four equations—Equation (3.11) written for $\alpha = w$ and $\alpha = n$, Equation (3.10), and Equation (3.12)—represent the governing equations for weakly compressible, immiscible two-fluid flow in porous media. They form a closed system, and can thus be solved for the four primary unknowns: p_w, p_n, s_w, and s_n. Such solutions require functional forms for the nonlinear relationships $\boldsymbol{k}_{r,\alpha}(s_\alpha)$ and $p^{cap}(s_w)$.

One of the interesting aspects of the equations for multiphase flow is the number of ways the equations can be combined and rewritten, while still representing the same system. We saw this to some extent in Chapter 2, where we wrote the same equation in terms of either hydraulic head or pressure. Because the multiphase equations involve more than one equation, and because those equations are nonlinear, there are significantly more opportunities to manipulate and rewrite the equations. We will consider a few examples of how these equations can be manipulated, and in the process introduce simplified equation forms that will be useful for the description of CO_2–brine systems.

Incompressible Fluids and Solid Matrix

We begin with a simple modification to Equation (3.11) based on the assumption of incompressibility. This means the compressibility is equal to zero, $c_\alpha = 0$, and the densities can be assumed to be constant. Equation (3.11) then takes the following form:

$$\phi \frac{\partial s_\alpha}{\partial t} - \nabla \cdot [\lambda_\alpha \boldsymbol{k}(\nabla p_\alpha - \rho_\alpha \boldsymbol{g})] = \phi \frac{\partial s_\alpha}{\partial t} + \nabla \cdot \boldsymbol{u}_\alpha = \upsilon_\alpha. \tag{3.13}$$

While not much different from Equation (3.11), we will find Equation (3.13) to be convenient to use and to be helpful to illustrate several important points. Note that we use the volumetric source term υ_α in Equation (3.13).

Summed Equations

While we can solve the two phase equations represented by Equation (3.11), it is often more convenient to replace one of those equations by the sum of the two equations. Prior to summing the equations, let us divide through by the density, ρ_α. The summed equations then take the following form:

$$\phi \sum_\alpha \frac{\partial s_\alpha}{\partial t} + \sum_\alpha \left(s_\alpha c_\alpha \frac{\partial p_\alpha}{\partial t} - \frac{1}{\rho_\alpha} \nabla \cdot [\rho_\alpha \boldsymbol{u}_\alpha] \right) = \sum_\alpha \upsilon_\alpha.$$

The first term is equal to zero because Equation (3.10) implies that $\frac{\partial}{\partial t}(s_w + s_n) = 0$. Therefore, the general summed equation is simply

$$\sum_\alpha s_\alpha c_\alpha \frac{\partial p_\alpha}{\partial t} - \frac{1}{\rho_\alpha} \nabla \cdot [\rho_\alpha \boldsymbol{u}_\alpha] = \sum_\alpha \upsilon_\alpha. \tag{3.14}$$

In the case of incompressible fluids and solid matrix, the system simplifies considerably because the compressibility terms are zero and the remaining densities cancel. With the total flux and the total source/sink term defined as $\boldsymbol{u}_\Sigma = \boldsymbol{u}_w + \boldsymbol{u}_n$ and $\upsilon_\Sigma = \upsilon_w + \upsilon_n$, Equation (3.14) takes the following form:

$$\nabla \cdot \boldsymbol{u}_\Sigma = \upsilon_\Sigma. \tag{3.15}$$

The simplicity of this final equation is one of the reasons that summing the phase equations is convenient. This summed equation is sometimes called the total pressure equation, or simply the pressure equation.

Shallow Air–Water Systems: Richards' Equation

Richards' equation is a simplified representation of a two-phase flow system that is often used to describe water movement in unsaturated soils. The approach is applicable to a two-fluid porous medium where the two fluids are water and air ($\alpha = w,a$), and the domain of interest is the shallow soil zone whose top boundary corresponds to the land surface. The important properties that allow for simplification are the following: (1) the viscosity of air is about two orders of magnitude less than the viscosity of water; (2) the density of air is about three orders of magnitude less than the density of water; and (3) the air phase often maintains a continuous pathway to the land surface, where the pressure is (by definition) atmospheric pressure. The low density of air means that within the soil zone, the density term $\rho_a gz$ will not cause significant changes in pressure in the vertical direction. The low viscosity of air means that it has a very high mobility (recall that viscosity is in the denominator of the mobility coefficient, λ_a), which means that air can usually move in the soil zone under very little gradient of air pressure. If the air has as its top boundary condition a fixed pressure equal to atmospheric pressure and it only builds up small pressure gradients, then the pressure of the air will be close to atmospheric pressure everywhere in the soil. Therefore, we will often assume that the air pressure can be well approximated by simply setting it equal to atmospheric pressure everywhere. That is, we assume air has infinite mobility and therefore requires vanishingly small pressure gradients to drive its movement. If the air pressure is assumed to be equal to atmospheric pressure everywhere, then we do not have to solve for p_a, meaning that one of the primary unknowns is eliminated. Therefore, one of the equations can be eliminated as well, and this equation is taken to be the air-phase equation.

Under these conditions, the capillary pressure is essentially the negative of the water pressure (the capillary pressure is positive because the water pressure is less than the atmospheric pressure in the unsaturated zone). In addition, it is customary in soil science to use the pressure head instead of the pressure, where pressure head is defined by $h_{pr} = p_w / \rho_w g$. Neglecting the higher order couplings from spatial and temporal derivatives of density and porosity, the phase equation for water derived from Equation (3.11) may be written as follows:

$$\frac{\partial(\phi s_w)}{\partial t} + \phi s_w S_s \frac{\partial h_{pr}}{\partial t} - \nabla \cdot [k_{r,w} \boldsymbol{\kappa} (\nabla h_{pr} + \nabla z)] = \upsilon_w, \tag{3.16}$$

where $S_s = c_f \rho_w g$ is the specific storativity term (refer to Eq. 2.21). Standard notation in soil science is to define the "moisture content," $\theta_w = \phi s_w$, which can be substituted into Equation (3.16). Given a functional relationship for relative permeability as a function of water saturation (or moisture content), and for water pressure head (i.e. capillary pressure head) as a function of saturation or moisture content, Equation (3.16) can be solved for the water pressure and the associated water saturation. Once the water pressure and saturation are determined, the air-phase saturation can be calculated via Equation (3.10).

Steady State: The Capillary Fringe

Recall that in Chapter 2, we considered the governing equation for flow in unconfined aquifers. For those aquifers, we identified the top boundary as the water table, which we defined as the surface along which the water pressure is equal to atmospheric pressure. We then assumed there was a "sharp interface" representing an abrupt change in water saturation. Given the concept of capillary pressure, the relationship between capillary pressure and saturation, and our discussion about the unsaturated zone, we can reexamine the region around the water table. In groundwater hydrology, the "capillary fringe" is often defined as the region where water fills all of the pore space (i.e., $s_w = 1$) but has pressure below atmospheric pressure. That is, it is a saturated zone above the water table, corresponding to the capillary pressure range between zero and the capillary entry pressure (for a drainage profile). We will use a more general definition of the term "capillary fringe" to denote the entire region of variable saturation, from the water table to the elevation where the water saturation is equal (or very close) to its residual value.

If we assume that the water pressure is close to a vertical equilibrium, then the water pressure changes in the vertical with a slope of $-\rho_w g$. If the air pressure above the water table is assumed to remain close to atmospheric pressure, then the capillary pressure moving upward from the water table is simply given by $p_a - p_w = p_{atm} - [p_{atm} - \rho_w g(z - \zeta_{WT})] = \rho_w g(z - \zeta_{WT})$, where $z = \zeta_{WT}$ denotes the elevation of the water table, which is defined here as the elevation where $p_w = p_{atm}$. To proceed, we use this expression for the pressure difference together with the (measured) capillary pressure curve:

$$p^{cap}(s_w) = p_a - p_w = \rho_w g(z - \zeta_{WT}).$$

We saw in Section 3.4 that the capillary pressure curve is a monotonically decreasing function of water saturation. Therefore, it is invertible. We call the inverse capillary-pressure function the *capillary-saturation function*, s_w^{cap}. We can then write

$$s_w(z - \zeta_{WT}) = s_w^{cap}(\rho_w g(z - \zeta_{WT})). \tag{3.17}$$

Equation (3.17) provides us with a simple expression for the saturation variation above the water table. Note that s_w^{cap} extends naturally such that below the water table we have $s_w^{cap}(z - \zeta_{WT} < 0) = 1$.

In the model for unconfined aquifers in Chapter 2, the addition of water to, or the subtraction of water from, the saturated zone of the aquifer was given by Equation (2.35), which related the rate of movement of the water table to the flux γ_t. If we think about the water table as a sharp interface, then the "drainable porosity" concept applies, and the resulting expression is that given in Equation (2.35). If we now consider the vertical movement of an equilibrium saturation profile like that of Figure 3.12, we see that the integrated volume of water that is either added to (rising water table) or subtracted from (falling water table) the saturated zone is simply equal to the drainable porosity, which is $\phi - \phi s_w^{res}$ as long as the domain is deep enough to allow the full saturation profile to exist. Therefore, while the distribution of water is different in the capillary fringe representation as compared to the sharp interface representation, the effect of a rising or falling water table is the same (assuming equilibrium profiles for water saturation). As such, a sharp interface representation of the water table as the top boundary of the saturated zone of an unconfined aquifer is often a reasonable approximation. This will not necessarily be true for the CO_2 problem, where the capillary fringe can play an important role in limiting horizontal flows through the fringe region (the unsaturated zone tends to be dominated by vertical flows and therefore, these horizontal flow effects are not important in the unsaturated zone while they can be very important in the CO_2 system).

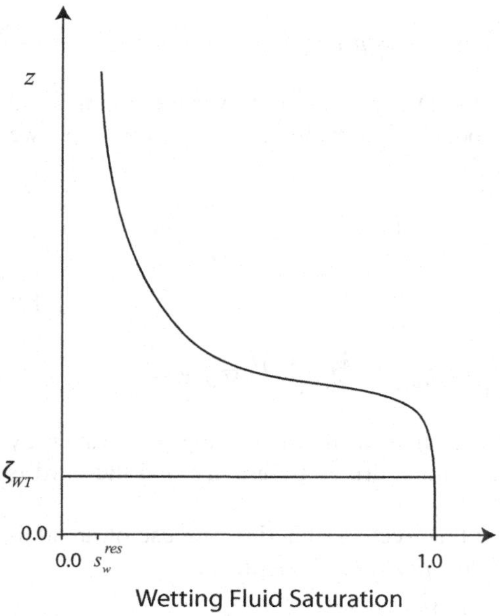

Figure 3.12 Typical saturation variation with depth near the water table.

3.6 PRACTICAL EQUATION FORMS AND SOLUTIONS

In this section we give some manipulations of the multiphase equations that are valuable in practice. Some of these manipulations involve a change of variables that allow for simpler numerical solution, while others give closed form solutions to special flow problems.

3.6.1 Average Pressure and Fractional Flow Formulations

The volumetric flux vector is given by Darcy's law, which has already been written for the multiphase case in Equation (3.2). We will now consider a number of different ways to rewrite this equation, based in large part on using a pressure, derived from the individual phase pressures, which can serve as an average pressure, or measure of potential, for the collective system of two fluid phases. We begin with the simplest possible definition of an average pressure for two phases—the simple arithmetic average, defined such that

$$\overline{p} = \frac{p_w + p_n}{2}. \tag{3.18}$$

With this definition of average pressure, the phase pressures may be written as follows:

$$p_w = \bar{p} - \frac{1}{2}p^{cap} \quad \text{and} \quad p_n = \bar{p} + \frac{1}{2}p^{cap}.$$

These pressures can be substituted into the Darcy equation to write the individual phase fluxes as well as the total flux. Introducing the total mobility $\lambda_\Sigma = \lambda_w + \lambda_n$, we can write the fluxes as

$$\begin{aligned} \boldsymbol{u}_w &= -\boldsymbol{\lambda}_w \boldsymbol{k}\left[\nabla\left(\bar{p} - \frac{1}{2}p^{cap}\right) - \rho_w \boldsymbol{g}\right] \\ \boldsymbol{u}_n &= -\boldsymbol{\lambda}_n \boldsymbol{k}\left[\nabla\left(\bar{p} + \frac{1}{2}p^{cap}\right) - \rho_n \boldsymbol{g}\right] \\ \boldsymbol{u}_\Sigma &= -\boldsymbol{\lambda}_\Sigma \boldsymbol{k}\nabla\bar{p} + (\rho_w \boldsymbol{\lambda}_w + \rho_n \boldsymbol{\lambda}_n)\boldsymbol{k}\boldsymbol{g} - \frac{(\boldsymbol{\lambda}_n - \boldsymbol{\lambda}_w)\boldsymbol{k}}{2}\nabla p^{cap}. \end{aligned} \tag{3.19}$$

Note that the total flow vector has three parts to it: the first is the flow driven by gradients of the average pressure, the second is driven by gravity, and the third is driven by gradients of capillary pressure.

Next, we write another form for the flux vectors, this time via use of an additional function known as the *fractional flow function*, $\boldsymbol{f}_\alpha$, defined as

$$\boldsymbol{f}_\alpha = \boldsymbol{\lambda}_\alpha \boldsymbol{\lambda}_\Sigma^{-1}.$$

For isotropic mobilities, the expression simplifies to $f_\alpha = \lambda_\alpha/\lambda_\Sigma$. With the fractional flow function, we can relate the phase flux vectors to the total flow vector with the following equations:

$$\begin{aligned} \boldsymbol{u}_w &= \boldsymbol{f}_w\left[\boldsymbol{u}_\Sigma + \boldsymbol{\lambda}_n \boldsymbol{k}\Delta_\alpha \rho \boldsymbol{g} + \boldsymbol{\lambda}_n \boldsymbol{k}\nabla p^{cap}\right] \\ \boldsymbol{u}_n &= \boldsymbol{f}_n\left[\boldsymbol{u}_\Sigma - \boldsymbol{\lambda}_w \boldsymbol{k}\Delta_\alpha \rho \boldsymbol{g} - \boldsymbol{\lambda}_w \boldsymbol{k}\nabla p^{cap}\right]. \end{aligned} \tag{3.20}$$

In this equation, the density difference $\Delta_\alpha\rho$ is defined as the difference between the wetting and nonwetting phase densities, that is, $\Delta_\alpha\rho = \rho_w - \rho_n$. Equation (3.20) allows the individual phase fluxes to be written as a fraction of the total flow, with modifications for both buoyancy and capillarity, thus eliminating the explicit presence of phase pressures.

3.6.2 The Buckley–Leverett Solution

In 1942, Buckley and Leverett published a solution for a simplified two-phase flow problem that has had remarkably widespread use. The solution is based on equations for one-dimensional flow of two incompressible fluids, where both density differences and capillary pressure effects can be ignored. If we ignore the formality of reducing dimensionality from three to one, and simply write the one-dimensional version of our equations, we can choose the summed phase equation as one of the governing equations and one of the phase equations as the second equation. The summed equation is the one-dimensional version of Equation (3.15), which, with no source or sink terms, and assuming the one spatial dimension is represented by x, can be written as

$$\frac{\partial u_\Sigma}{\partial x} = 0.$$

This implies that the total flow, u_Σ, is constant in space. Next, consider the equation for one of the phases, say the wetting phase. We use the expression for the wetting phase flux given by Equation (3.20). Substitution of Equation (3.20) into Equation (3.13), neglecting density and capillary effects, and remembering that u_Σ does not depend on x, yields the following equation that governs the movement of the wetting phase:

$$\phi \frac{\partial s_w}{\partial t} + u_\Sigma \frac{\partial f_w}{\partial x} = 0. \tag{3.21}$$

In Chapter 2, we noted in passing (Section 2.3.2) that two special cases of the single-phase flow equation corresponded to classical forms of partial differential equations, which we identified as parabolic (transient single-phase flow) and elliptic (steady-state flow). Equation (3.21) provides a model of another classic form of partial differential equation: a first-order hyperbolic equation. Solution of these types of equations are often based on so-called characteristic methods, which depend on the specific form of the fractional flow function f_w, which in turn depends on the functional forms for the relative permeability functions. To illustrate a simple version of the Buckley–Leverett solution, let us define nonlinear relative permeability functions given by a cubic polynomial as $k_{r,\alpha} = (s_\alpha)^3$, where for simplicity we ignore residual saturations. In this case, the fractional flow functions take the following form:

$$f_w = \frac{(s_w)^3}{(s_w)^3 + \frac{\mu_w}{\mu_n}(s_n)^3} \quad \text{and} \quad f_n = \frac{(s_n)^3}{\frac{\mu_n}{\mu_w}(s_w)^3 + (s_n)^3}.$$

For the case of equal viscosities, these curves are symmetric about $s_w = s_n = 0.5$ point. This symmetry is lost whenever the viscosity ratio of the two fluids is not equal to one. If we now return to Equation (3.21), we can use the curve for f_w to construct a solution to the equation.

If we expand the spatial derivative in Equation (3.21) using the chain rule of differentiation, we can write the following:

$$\phi \frac{\partial s_w}{\partial t} + u_\Sigma \frac{df_w}{ds_w} \frac{\partial s_w}{\partial x} = 0. \tag{3.22}$$

After division by porosity, we see that the expression in Equation (3.22) can be interpreted as a total derivative with respect to time (also known as a material derivative), where the total derivative is measured in a specific moving coordinate system. That is, we can relate the total derivative to the partial derivatives by the expansion

$$\frac{Ds_w}{Dt} \equiv \frac{\partial s_w}{\partial t} + \frac{dx}{dt}\frac{\partial s_w}{\partial x} = \frac{\partial s_w}{\partial t} + v\frac{\partial s_w}{\partial x}$$

where v is the velocity of motion, and the total derivative is the change in s_w measured in that moving coordinate system. Equation (3.22) indicates that if the velocity of observation is defined by $(u_\Sigma/\phi)(df_w/ds_w)$, then the equation satisfied by the saturation s_w will be $Ds_w/Dt = 0$. That is, in this moving coordinate, the saturation does not change. This is the idea that allows the solution to be constructed.

This somewhat heuristic argument is essentially the underlying structure for so-called characteristic methods of solution, which can form the basis for both analytical and numerical solutions. In this case, the solution is analytical, but it has one complicating feature: the velocity depends on the derivative of the fractional flow curve, which is a nonlinear function of the saturation s_w. As seen in Figure 3.13a, part of the curve is concave upward, and this causes problems. For example, assume we begin with a saturation profile in space that is a linear variation from one to zero over some interval of the x-axis, as shown in Figure 3.13b, and we keep the saturation fixed at the left boundary. Consider two specific points on the fractional flow function, say at $s_w = 0.2$ and $s_w = 0.4$. The slope at $s_w = 0.4$ is larger than the slope at $s_w = 0.2$, as can be seen from Figure 3.13a. Therefore, the speed with which the (constant) value of saturation, $s_w = 0.4$, moves is greater than the speed for the saturation of $s_w = 0.2$. Although the initial distribution of saturation (Figure 3.13b) has the lower saturation ahead of the higher saturation on the x-axis, the larger speed of the higher saturation means that it will eventually overtake the lower saturation, leading to a multi-valued solution for saturation. The resulting profile is indicated in Figure 3.14 by the multi-valued dashed curve showing saturation as a function of space. This is obviously a nonphysical solution to the problem, and the physically

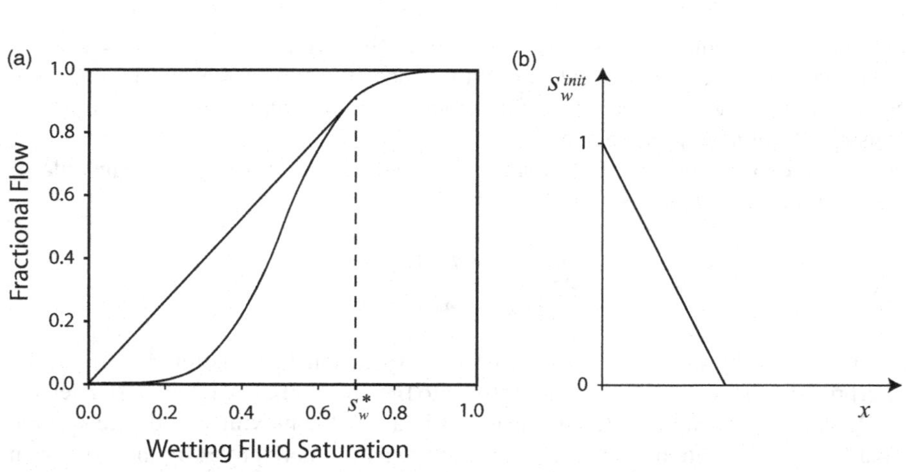

Figure 3.13 Fractional flow function with its convex hull (a), and an initial condition that develops into a shock (b).

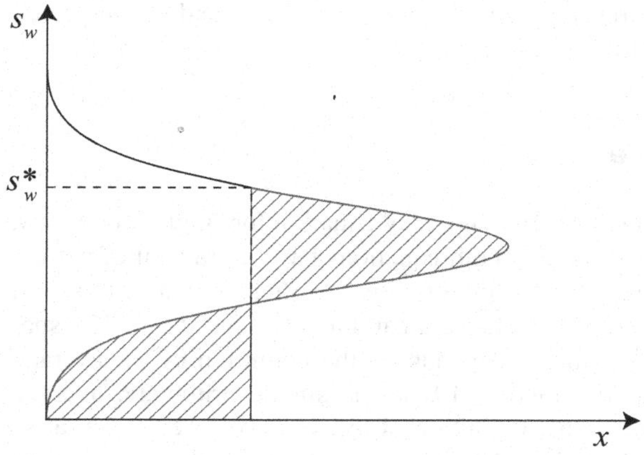

Figure 3.14 A typical Buckley–Leverett solution displaying a shock followed by a rarefaction wave.

motivated constraint that the solution should be single-valued must be imposed. The appropriate single-valued function can be derived in several ways—for example, using the general theory of hyperbolic equations, or using an entropy constraint for the system. The resulting single-valued function contains a discontinuity (a shock) and has the property that the amount of fluid mass in the system at any given time is the same as the mass associated with the multi-valued function. That is, the shaded areas on either side of the vertical line denoting the shock in Figure 3.14 must be the same. It turns out that the magnitude of the shock (i.e., the size of the jump in saturation) and its speed can be determined by constructing a line that hits the tangent point along the fractional flow curve, as shown in Figure 3.13a. The slope of the straight line is the value of df_w/ds_w used to calculate the velocity of the shock, and the value of s_w^* in Figure 3.13a gives the magnitude of the shock. For all saturations greater than this value, the solution spreads out instead of sharpening, in what is referred to as a *rarefaction wave*. The slope of the fractional flow function in this range of saturations can be used directly to determine the appropriate velocities with which to move the saturations. So, by direct construction, the solution to this problem consists of a shock wave and a rarefaction wave, as illustrated in Figure 3.14. This represents the classical Buckley–Leverett solution, which is an important analytical solution for two-phase flow in porous media.

It is also instructive to consider the case of relative permeability functions that are linear functions of saturation, for example, $k_{r,\alpha} = s_\alpha$. Now, for the case of equal fluid viscosities, the fractional flow functions are simply linear, that is, $f_\alpha = s_\alpha$, and the governing equation reduces to a simple linear first-order hyperbolic equation given by $\phi(\partial s_w/\partial t) + u_\Sigma(\partial s_w/\partial x) = 0$. This equation takes any initial condition for saturation and propagates it, unchanged, along the x-axis with constant speed given by u_Σ/ϕ. This is a useful result to remember: when relative permeability functions are linear and the viscosity ratio is equal to one, the transport equation becomes linear such that the solution is transported with constant velocity, and neither shock waves nor rarefaction waves are created.

3.6.3 Global Pressure

If we compare Equation (3.19) to the original Darcy equation for single-phase flow, for example Equation (2.10), we see that both equations have a gradient of a pressure term and a gravity term, albeit with different coefficients. In contrast to Equation (2.10), Equation (3.19) also includes a capillary pressure term. The specific form of Equation (3.19) obviously depends on the definition of the average pressure—the average used in Equation (3.19) is the simple arithmetic average given by Equation (3.18). Because the definition of average pressure was an arbitrary choice, we might consider other definitions. It turns out that if capillary pressure and relative permeabilities are not functions of space, and if relative permeability is taken as a scalar function, we can define a specific average pressure, called the *global pressure*, p_G, which has the property that when used to define the total flow vector u_Σ, the resulting expression contains only a gradient of the global pressure and a gravity term. That is, the use of the global pressure makes the total flow vector for two-phase flow have a form analogous to Darcy's equation for single-phase flow. This can sometimes have advantages when solving the system of two-phase flow equations, as will be seen later in the book with calculations in Chapter 5.

The global pressure has the following definition for scalar mobilities:

$$p_G \equiv \overline{p} + \int_{s_{w,0}}^{s_w} \left(\frac{1}{2} - f_w \right) \frac{dp^{cap}}{ds'_w} ds'_w, \tag{3.23}$$

where $s_{w,0}$ is defined as the wetting saturation where $p^{cap}(s_{w,0}) = 0$. When this is substituted into Equation (3.19), the result is

$$\boldsymbol{u}_\Sigma = -\lambda_\Sigma \boldsymbol{k} \nabla p_G - (\rho_w \lambda_w + \rho_n \lambda_n) \boldsymbol{k} \boldsymbol{g}. \tag{3.24}$$

Equation (3.24) relates the total flow to the gradient of the global pressure and gravity. As compared to Equation (3.19), it has no explicit representation of capillary pressure in the equation.

We will refer to the global pressure again later, so for completeness we remark that the definition is skew-symmetric in terms of wetting and nonwetting phase (see Box 3.2). As with all uses of pressure for incompressible problems, the choice of datum is also somewhat arbitrary. This is evidenced by Equation (3.24) being valid independently of the lower integration limit in the definition of global pressure (as long as it is constant in space). Therefore, the global pressure is often specified without recourse to a datum or lower integration limit, and in terms of one of the phase pressures, for example,

$$p_G = p_w + \int^{s_n} f_n \frac{dp^{cap}}{ds'_n} ds'_n. \tag{3.25}$$

BOX 3.2 *Existence of a Global Pressure*

Global pressure expresses the idea that the nongravitational component of the total flux $\boldsymbol{u}_\Sigma$ can be written as the gradient of a potential. This idea is more important than the explicit definition given in Equation (3.23) or (3.25), as it allows for a simpler expression for total flux which has a weaker dependence on the saturation distribution. In other words, returning to the expression for total flux in Equation (3.19), and neglecting gravity with isotropic fluid mobilities, we ask if a global pressure p_G exists such that

$$\frac{\boldsymbol{k}^{-1}}{\lambda_\Sigma}\boldsymbol{u}_\Sigma = -\nabla p_G = -\nabla \bar{p} - \frac{(\lambda_n - \lambda_w)}{2\lambda_\Sigma}\nabla p^{cap}.$$

From calculus, we know that a vector function can be written as a gradient of a potential if and only if the curl of the vector function is zero. Therefore, if p_G exists then $\nabla \times \nabla p_G = 0$. Taking the curl of the above expression, we obtain from the last equality that

$$\nabla \times ((1-2f_w)\nabla p^{cap}) = \mathbf{0}$$

must hold for a global pressure to exist. Here we have rearranged the expression in the parenthesis to get the fractional flow instead of the difference in mobilities. If the capillary-pressure and fractional flow functions are functions of saturation only (and not heterogeneous in space), we see that

$$\begin{aligned}\nabla \times ((1-2f_w)\nabla p^{cap}) &= \nabla \times \left((1-2f_w)\frac{dp^{cap}}{ds_w}\nabla s_w \right) \\ &= \frac{d}{ds_w}\left((1-2f_w)\frac{dp^{cap}}{ds_w} \right)\nabla s_w \times \nabla s_w + \left((1-2f_w)\frac{dp^{cap}}{ds_w} \right)\nabla \times \nabla s_w \\ &= \mathbf{0}\end{aligned}$$

since both $\nabla s_w \times \nabla s_w$ and $\nabla \times \nabla s_w$ are evaluated as the zero vector. However, if either constitutive function is spatially heterogeneous, then $\nabla \times ((1 - 2f_w)\nabla p^{cap}) \neq 0$, and the global pressure, in the sense of a potential driving the total flow, does not exist. However, the definition in Equation (3.25) may still be evaluated, and this pressure, which we will still denote as the global pressure, remains a good choice for an independent variable when solving the multiphase flow equations numerically.

3.7 REDUCTION OF DIMENSIONALITY

In Chapter 2 we introduced the idea of reduction of dimensionality, wherein the three-dimensional governing partial differential equations were reduced to two-dimensional equations for cases where vertical flow within the formation of interest could be neglected. This assumption is not applied nearly as widely for multi-fluid systems, in part because the nonlinearities associated with the fluid saturations make the overall flow system much more complex. However, there are cases where reduction of dimensionality can be applied, and in such cases, it can often lead to substantial benefits in terms of system simplification. One such case is when the two fluid phases involved have strong density differences, such that $\Delta_\alpha \rho$ is large. In these cases, the fluids tend to segregate by density, with the lighter fluid stratified above the denser fluid. Depending on the specific material properties, in particular the capillary pressure function, the transition zone between the two fluids can be quite thin and, therefore, in some cases, it may be reasonably approximated by a sharp interface. For simplicity, we assume a sharp interface in this section. In Chapter 4 we will relax this assumption and include a finite transition zone based on the relationship between saturation and capillary pressure.

Within each of the two zones, an assumption of vertical equilibrium defines the vertical distribution of fluid pressures and allows each of the fluid zones to be vertically integrated. The resulting description is suitably simplified. In this section we outline this procedure, applied to stratified flows separated by large density contrast. While more specific discussion of what "large" density contrast means will take place in Chapter 4, for now we recall that the density differences between brine and supercritical CO_2 tend to be at least several hundred kilograms per cubic meter, which is quite large (compare this to seawater and freshwater, which differ by only 25 kg/m^3 but still show significant density segregation). Therefore, the CO_2–brine system is an excellent candidate for these kinds of simplifications.

Consider the formation illustrated in Figure 3.15, where the bottom of the formation is at elevation ζ_B, the top is at elevation ζ_T, and two fluids exist within the formation, separated by a surface whose elevation is denoted by ζ_I. If this were a CO_2 injection scenario, the lighter fluid (CO_2) would be displacing the resident denser fluid (brine), at least during the injection period. Because our interest is in the CO_2–brine problem, we will denote the two fluids by $\alpha = b$ for the brine phase and $\alpha = c$ for the CO_2 phase. As we saw in Section 3.4, residual saturations occur upon displacement of one fluid by another. So we can assume the CO_2 region has some of its pore space filled with brine, which exists at its residual saturation and therefore is immobile ($k_{r,b} = 0$ when $s_b \le s_b^{res}$). Therefore, the system is one for

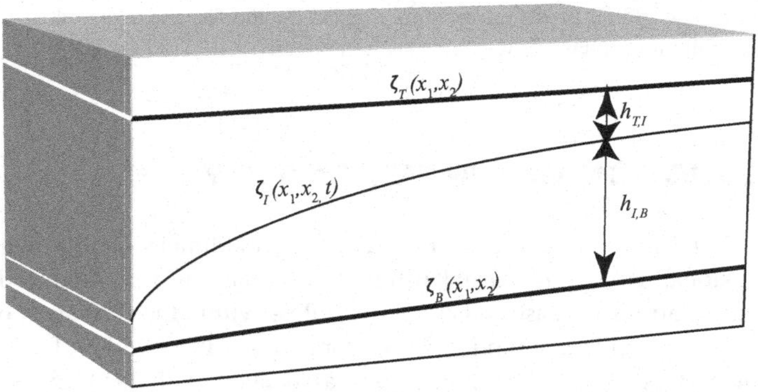

Figure 3.15 Sketch of an aquifer with the various interfaces used in the vertical integrations.

which each of the two regions has one mobile fluid phase; in the bottom region $s_b = 1$, $s_c = 0$; and in the top region $s_b = s_b^{res}$, $s_c = 1 - s_b^{res}$. For now, we will assume that the top and bottom formation boundaries align in general with the horizontal direction, so that the coordinate direction x_3 coincides with the vertical direction z. We will also assume incompressibility for the fluids and solid. In this case, flow in each of the two regions is equivalent to single-phase flow, with a reduction in the apparent porosity in the top region by the factor $1 - s_b^{res}$ (and the associated reduction in permeability). Therefore, the vertically integrated equations for the single-phase case, as derived in Chapter 2, can be used directly here. We only need to realize that the interface ζ_I is now the bottom boundary for the top region, and the top boundary for the bottom region. We also need to write a condition for the flux across that boundary that is analogous to the equation used for the water table in Chapter 2, which is Equation (2.35). And, we need to recognize that the assumption of vertical equilibrium now must take into account the interface boundary and the two fluid densities, so that the vertical equilibrium equation for pressure becomes the following:

$$p(x_1, x_2, z, t) = \begin{cases} p_B(x_1, x_2, t) - \rho_b g[z - \zeta_B(x_1, x_2)], & \zeta_B \le z \le \zeta_I \\ p_B(x_1, x_2, t) - \rho_b g[\zeta_I(x_1, x_2, t) - \zeta_B(x_1, x_2)] - \rho_c g[z - \zeta_I(x_1, x_2, t)], & \zeta_I < z \le \zeta_T \end{cases} \tag{3.26}$$

where we have ignored capillary entry pressure at the interface. The integrated equation for the bottom (brine) region is then given by the following (note that in the brine region, $\partial s_b / \partial t = 0$ and we assume incompressibility):

$$\int_{\zeta_B}^{\zeta_I} \nabla \cdot \boldsymbol{u}_b \, dz = \nabla_{\|} \cdot \int_{\zeta_B}^{\zeta_I} \boldsymbol{u}_{b,\|} dz + \boldsymbol{u}_{b,I} \cdot (\boldsymbol{e}_z - \nabla_{\|} \zeta_I) - \boldsymbol{u}_{b,B} \cdot (\boldsymbol{e}_z - \nabla_{\|} \zeta_B) = \int_{\zeta_B}^{\zeta_I} \upsilon_b dz = \Upsilon_b.$$

From this we define coarse-scale variables as in Section 2.3.1 and write

$$\nabla_{\|} \cdot \boldsymbol{U}_b + \boldsymbol{u}_{b,I} \cdot (\boldsymbol{e}_z - \nabla_{\|} \zeta_I) - \Upsilon_B = \Upsilon_b. \tag{3.27}$$

We define the coarse flux in terms of coarse-scale permeability and mobilities:

$$\boldsymbol{U}_b \equiv \int_{\zeta_B}^{\zeta_I} \boldsymbol{u}_b \, dz = -\boldsymbol{\Lambda}_b \boldsymbol{K} (\nabla p_B + \rho_b g \nabla \zeta_B), \tag{3.28}$$

where the coarse mobility term arises from the integrated coefficient given by $\int_{\zeta_B}^{\zeta_I} \boldsymbol{\lambda}_{b,\|} \boldsymbol{k}_{\|} dz$, which we replace by the product of the integrated permeability, $\boldsymbol{K}$, and the vertically averaged mobility, denoted by $\boldsymbol{\Lambda}_b \equiv \int_{\zeta_B}^{\zeta_I} \boldsymbol{\lambda}_{b,\|} \boldsymbol{k}_{\|} dz \, \boldsymbol{K}^{-1}$. Note that both $\boldsymbol{K}$ and $\boldsymbol{\Lambda}_b$ are 2×2 matrices (or "tensors"), and their product is another 2×2 matrix. Note also that the integrated mobility $\boldsymbol{\Lambda}_\alpha$ may be anisotropic even if the fine-scale mobility is a scalar, isotropic function. In Equation (3.28), we have also used the observation that Equation (3.26) implies

$$\nabla_{\|} p_b(x_1, x_2, \zeta_B \le z < \zeta_I, t) = \nabla p_B + \rho_b g \nabla \zeta_B.$$

Use of the interface condition, Equation (2.35), with no flux through the interface ($u_t = 0$), and with the drainable porosity given by $\phi_d = \phi(1 - s_b^{res})$ to take into account only that part of the pore space available for the CO_2 phase, we obtain from Equation (3.27) the following equation:

$$\phi(1 - s_b^{res}) \frac{\partial \zeta_I}{\partial t} + \nabla \cdot \boldsymbol{U}_b = \Upsilon_b + \Upsilon_B. \tag{3.29}$$

Similarly, we can integrate over the top region, using Equation (3.26) to project bottom pressure into the top region. We find the following equation,

$$-\phi(1 - s_b^{res}) \frac{\partial \zeta_I}{\partial t} + \nabla \cdot \boldsymbol{U}_c = \Upsilon_c - \Upsilon_T, \tag{3.30}$$

where we have defined the coarse-scale CO_2 flux similarly to Equation (3.28) as

$$\boldsymbol{U}_c \equiv -\boldsymbol{\Lambda}_c \boldsymbol{K}(\nabla p_B - \Delta_\alpha \rho g \nabla \zeta_I + \rho_b g \nabla \zeta_B).$$

There are several details to note here, two of which we will mention. The first is that we can relate the position of the interface, ζ_I, to the thickness of the CO_2 and brine layers. We have already defined the thickness of the formation in Chapter 2 as $H \equiv \zeta_T - \zeta_B$. However, we now have other thickness measures, which we choose to define in a consistent way through the use of the variable h with a double subscript to denote the top and bottom boundaries of the region of interest. For example, we define the thickness of the CO_2 and brine regions as

$$h_{T,I} \equiv \zeta_T - \zeta_I \quad \text{and} \quad h_{I,B} \equiv \zeta_I - \zeta_B.$$

Note that we can also redefine the overall formation thickness as $H = h_{T,B} \equiv \zeta_T - \zeta_B$. We can now substitute appropriate thicknesses in Equations (3.29) and (3.30) in place of the interface height, so that the equations are written with the two primary unknowns being the bottom pressure and the CO_2 thickness:

$$-\phi(1 - s_b^{res}) \frac{\partial h_{T,I}}{\partial t} - \nabla \cdot [\boldsymbol{\Lambda}_b \boldsymbol{K}(\nabla p_B + \rho_b g \nabla \zeta_B)] = \Upsilon_b + \Upsilon_B$$

$$\phi(1 - s_b^{res}) \frac{\partial h_{T,I}}{\partial t} - \nabla \cdot [\boldsymbol{\Lambda}_c \boldsymbol{K}(\nabla p_B + \Delta_\alpha \rho g \nabla h_{T,I} + \rho_c g \nabla \zeta_T - \rho_b g \nabla h_{T,B})] = \Upsilon_c - \Upsilon_T. \tag{3.31}$$

Here we have used that

$$\frac{\partial h_{I,B}}{\partial t} = \frac{\partial (h_{T,B} - h_{T,I})}{\partial t} = -\frac{\partial h_{T,I}}{\partial t}.$$

Note that for the simplified case of horizontal and flat top and bottom boundary surfaces, and with no external sources or sinks and no leakage through the top or bottom boundaries, the equations simplify to the following form:

$$
\begin{aligned}
-\phi(1-s_b^{res})\frac{\partial h_{T,I}}{\partial t} - \nabla\cdot[\Lambda_b K \nabla p_B] = 0 \\
\phi(1-s_b^{res})\frac{\partial h_{T,I}}{\partial t} - \nabla\cdot[\Lambda_c K(\nabla p_B + \Delta_\alpha \rho g \nabla h_{T,I})] = 0.
\end{aligned}
\tag{3.32}
$$

Note that these equations can be summed, as was performed in Section 3.5.2, to derive a total pressure equation. The manipulations discussed in Section 3.6 can then be applied to this system, written in terms of a pressure equation and one saturation equation.

Second, we recall that the vertically integrated mobility terms Λ_α are defined as

$$
\Lambda_\alpha \equiv \int_{\zeta_B}^{\zeta_I} \lambda_{\alpha,\parallel} k_\parallel dz\, K^{-1}
$$

These mobilities are close to linear functions of the interface height ζ_I (or equivalently, the thickness of the CO_2 region, $h_{T,I}$), varying between 0 and $\lambda_\alpha(1 - s_{r,\alpha})$. To see this, consider the case of a homogeneous medium with isotropic mobility on the fine scale, where we take the brine phase as an example:

$$
\Lambda_b = \int_{\zeta_B}^{\zeta_I} \lambda_b\, dz k (k h_{T,B})^{-1} = \lambda_b(1)\frac{h_{I,B}}{h_{T,B}}.
$$

This is a very useful simplification when considering homogeneous, vertically integrated sharp interface models.

Equations (3.32) form a set of two nonlinear equations for the two primary unknowns, $h_{T,I}$ and p_B. We will revisit this set of equations in Chapters 4 and 5, where solutions to the equations, including analytical solutions, will be presented and discussed.

3.8 NUMERICAL SOLUTIONS

In Section 2.6 we considered a few basic ideas related to numerical solutions for single-phase flow. These involved simple discretizations in both space and time. The spatial domain was broken into subdomains within which local mass balance constraints were enforced, with fluxes approximated across the boundaries of each of the subdomains or grid cells. The time domain was discretized and simple time-stepping algorithms were used involving two time levels, one where the solution was already known and the second where the solution was to be computed. Euler time-stepping methods were introduced, both explicit and implicit. We noted that implicit methods are usually applied for the single-phase flow equations, with a matrix solution at each time step providing the discrete values for either pressure or hydraulic head.

In the present case of two-fluid flows in porous media, the system is noticeably more complex. We now have multiple, coupled equations to solve, and the equations are nonlinear. This requires additional considerations, including which equation set to solve, how or if implicit and explicit methods can be combined, and how to treat the nonlinearities in the implicit solutions. It turns out that standard procedures have evolved for the solution of the two-phase flow equations, and these procedures have proven to be effective in practice. We will outline two of those procedures here, focusing on the two-dimensional vertically integrated forms of the governing equations. The basic ideas apply equally well to the full three-dimensional equations.

We begin with the choice of equations to solve and the primary unknowns. Almost all traditional numerical solution methods solve for at least one value of pressure. The specific pressure might be one of the phase pressures, the average pressure defined in Equation (3.18) or the global pressure in Equation (3.23). Typically, the second primary variable is one of the saturations. Regarding specific forms of the governing equations, it is almost always beneficial to choose as one of the equations the summed phase equations, for example, Equation (3.14). As mentioned earlier, this equation is often referred to as the *total pressure equation*, or just the *pressure equation,* because it is usually associated with solution for the pressure variable. Note that when the system is incompressible, saturation does not appear explicitly in the equation; rather, it appears only through the nonlinear dependencies in the phase mobilities (and capillary pressure, if the global pressure is not chosen). If the summed phase equation is one of the equations to be solved, then one of the phase equations is the second equation. Coupled with the appropriate functional relationships between relative permeability and saturation, and between capillary pressure and saturation, the equation set can be solved.

One common solution procedure is referred to as the IMplicit Pressure Explicit Saturation, or IMPES, method. Assume a simple spatial and temporal discretization like that used in Section 2.6. On this discretized domain, the IMPES method proceeds by solving the pressure equation implicitly for the pressure at the new time level. The calculation uses values for the nonlinear coefficients in the pressure equation based on saturation values from the previous time step. So the pressure is evaluated implicitly but the coefficients are not; instead, they are lagged in time to linearize the equation. This new pressure is used to calculate updated fluxes across all grid–cell boundaries, and with these updated fluxes the discrete version of the divergence terms $\boldsymbol{\nabla}_{\|} \cdot \boldsymbol{U}_\alpha$ is calculated. Any non-zero top or bottom fluxes and source or sink terms are also included. All of these fluxes are taken as known information, and in the phase equations to be solved, the saturation time derivative is discretized and the saturation at the new time level becomes the only unknown in the equation. Therefore, the saturation calculation is performed explicitly. With newly updated saturations, the nonlinear coefficients in the equations are updated, including capillary pressure terms, and the algorithm proceeds to the next time step. The implicit calculation of pressure and the explicit calculation of saturation leads to the name of the method.

For particularly difficult problems, the explicit nature of the IMPES approach sometimes fails to give satisfactory solutions. In such cases, it may be necessary to use a so-called *fully implicit* method. As the name indicates, this method uses numerical approximations for which all terms are evaluated at the new (unknown) time level. This requires solution of a coupled set of nonlinear algebraic equations, resulting from the discrete equations analogous to those in Equations (2.56) and (2.57). With this set of nonlinear algebraic equations, any number of standard nonlinear solution techniques can be used. The most common is the Newton–Raphson method and variants thereof.

Another aspect of the multiphase flow case that deserves mention is the evaluation of coefficients at the grid–cell boundaries. In the pressure equation case, we argued that the harmonic average was a natural choice for the transmissivity coefficient, using the analogy of flow in layered media in the direction orthogonal to the layering. For the saturation equation, the harmonic average turns out to be a bad choice. This is most easily seen when a grid cell has no saturation of one of the phases at a certain time, but that phase appears in a neighboring cell. Given a saturation of zero in the cell of interest, the relative permeability within that cell will also be zero, and therefore, the mobility will be zero. The harmonic average of any pair of values is always zero when one of them is zero. Therefore, if the harmonic average is assigned to the cell edges, the phase can never enter that cell. Instead of the harmonic average, the mobilities are either averaged using a variant of the arithmetic average, or often they are "upstream weighted" by assigning the value to an interface from the cell that is in the upstream flow direction. These strategies become more complex for anisotropic mobilities since they disalign the phase and total velocities, making determination of upstream directions less well defined. This is part of the explanation of why most numerical implementations only allow for relative permeability to be a scalar.

3.9 COMPRESSION AND RECONSTRUCTION

Recall that in Section 2.7 we introduced the concept of compression and reconstruction operators, wherein detailed information at a particular scale of resolution, called the fine scale, is replaced by some averaged measure that is defined at a larger, or coarse scale. That is, information from the fine scale was compressed into an averaged, and therefore much less detailed, measure. From the compressed information, the reconstruction operator acted to try to reproduce the fine-scale structure of the original data. We used pore-scale geometry and its relationship to porosity and permeability as one example, and the transformation from hydraulic conductivity to transmissivity and then back again as a second example. In this section we again apply these concepts to an example, which generalizes directly to the CO_2–brine problem, as we will see in the subsequent chapters.

We return to the case of homogeneous unconfined aquifers, and the water table. We consider the simplest description of the vertical distribution of fluids. As such, the coarse variable may be a measure of total fluid content ζ (measured in units of length or volume per area), and the associated compression operator from the fine scale can be identified as integration, $\zeta = Cs_w \equiv \int_{\zeta_B}^{\zeta_T} s_w dz + \zeta_B$. We recognize that for this homogeneous system, ζ is a coarse-scale variable closely related to the height of the water table, ζ_{WT}. Note that even though we are thinking of an unconfined aquifer, we have specified an upper limit of integration ζ_T which we take to be the land surface. For simplicity, we will in the following assume that the datum for height is chosen at the bottom aquifer boundary, $\zeta_B = 0$.

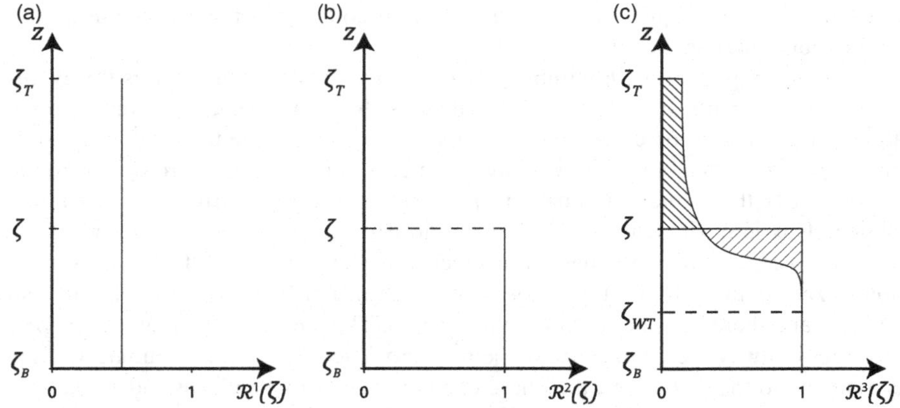

Figure 3.16 Various fine-scale vertical reconstructions of the same coarse-scale saturation: (a) constant, (b) sharp interface, (c) and capillary fringe.

The challenge will often be to reconstruct from the coarse-scale saturation a fine-scale distribution of saturations in the vertical dimension. To illustrate the role of our understanding of the subscale physics on the reconstruction, we consider three example reconstruction concepts. For each reconstruction, a given value of ζ is converted to a fine-scale representation that has the important property that it preserves the (coarse-scale) value. The three options are: (1) uniform distribution in z, (2) sharp interface representation, and (3) capillary fringe representation. Each of these is shown schematically in Figure 3.16.

The first reconstruction operation is simple: we let the reconstructed fine-scale saturation be independent of z and equal to the average saturation:

$$\widehat{s_w}(z) = \mathcal{R}^1\zeta \equiv \zeta / h_{T,B}.$$

Recall from Chapter 2 that we use $\mathcal{R}$ to represent reconstruction operators and hats to represent reconstructed variables. We also use superscript 1 to denote the first (of three) reconstruction choices. For this reconstruction, there is no vertical structure, and while the reconstruction satisfies the constraint $\zeta = C\widehat{s_w}$, we immediately recognize that it does not embody our understanding of the subscale physics, since it does not include the effect of gravity segregating the two fluids.

The second option uses the sharp interface assumption. If we assume that the aquifer is completely saturated by water below the water table, and that the soil is dry above the water table, then we can define the reconstruction

$$\widehat{s_w}(z) = \mathcal{R}^2\zeta \equiv \begin{cases} 1 & \text{if} \quad z \le \zeta_{wT} \\ 0 & \text{if} \quad z > \zeta_{wT} \end{cases}.$$

Here, ζ is exactly equal to the saturated thickness of the unconfined aquifer. This reconstruction also satisfies $\zeta = C\widehat{s_w}$, and is closer to our physical understanding of the system since it is consistent with gravity segregation. However, it does not honor the capillary pressure–saturation relationship, except in the special case where the capillary saturation curve is given by the Heaviside step function:

$$s_w = s_w^{cap}(p_a - p_w) = \begin{cases} 1 & \text{if} \quad p_a - p_w \leq p_{entry}^{cap} \\ 0 & \text{if} \quad p_a - p_w > p_{entry}^{cap} \end{cases}.$$

This approximation is known as a delta soil in hydrology, and leads to several useful solutions including the widely used Green and Ampt solution. Note that use of the water table as the boundary in the saturation reconstruction implies that the entry pressure is zero; a non-zero entry pressure would lead to a saturated zone above the water table.

Finally, the third option involves reconstruction using the capillary fringe. In Equation (3.17) we found how the saturation varies above the water table ζ_{WT}. However, we still need to honor the constraint

$$\zeta = \mathcal{C}(\mathcal{R}^3\zeta) = \int_{\zeta_B}^{\zeta_T} s_w^{cap}(\rho_w g(z - \zeta_{WT}))dz.$$

This is a nonlinear equation for ζ_{WT}, which essentially defines the functional relationship $\zeta_{WT} = \zeta_{WT}(\zeta)$. Given this functional form, we can state the capillary fringe-based reconstruction as

$$\widehat{s_w}(z) = \mathcal{R}^3\zeta \equiv s_w^{cap}(\rho_w g[z - \zeta_{WT}(\zeta)]).$$

This third reconstruction now honors the balance between gravitational and capillary forces. It also gives us more understanding of the two other reconstructions, since we see that they appear as limiting cases when either gravity is negligible (leading to the first reconstruction), or when the variation in capillary forces are negligible (leading to the second reconstruction).

In general, we expect the best reconstruction to lie somewhere between the sharp interface model and the capillary fringe model. We note that any standard numerical simulator with insufficient grid resolution applied to the original three-dimensional equations will, by default, use the uniform reconstruction because each (relatively large) grid cell will approximate the saturation using a constant value over the cell. This highlights the importance of modeling these processes using numerical discretizations that are compatible with the scale at which the governing equations are written. These kinds of observations provide the motivation for multiscale modeling concepts presented in the subsequent chapters.

3.10 QUALITATIVE DESCRIPTION OF CO_2 STORAGE

With all of the information presented in the first nine sections of this chapter, we now try to gain a qualitative understanding of the multiphase flow aspects for the CO_2 storage problem. To do this, we will consider the timeline of injection into a deep permeable formation that is bounded above and below by low-permeability caprock formations and is initially filled with brine. We will assume injection from a single vertical well, although the processes that govern the system are essentially independent of the orientation of the injection well or wells. We also assume that the brine is the wetting fluid, and the injected CO_2 is the nonwetting fluid. Therefore, displacement of brine by CO_2 is a drainage process, while displacement of CO_2 by brine is an imbibition process.

Injection of CO_2 into the formation is achieved by raising the pressure of the CO_2 in the wellbore to a value above the entry pressure for the formation material. Because the field operator will typically want to inject as much CO_2 as possible, the injection pressure is likely to be much greater than the entry pressure, limited by requirements to remain below the estimated fracture pressure of the formation and caprock. At the pore scale, fluid–fluid interfaces that separate CO_2 and brine move into the formation as the injection is carried out. In the immediate neighborhood of these interfaces, components can transfer some of their mass from one phase to another—for example, CO_2 will dissolve into the brine across these interfaces. The concentration of dissolved CO_2 is usually taken to be at its equilibrium value in the brine immediately adjacent to the fluid–fluid interface. This dissolution will continue to take place until diffusion spreads the aqueous concentration within the pore space and eventually all of the brine in the vicinity of a fluid–fluid interface reaches the equilibrium concentration. At the same time, H_2O from the aqueous phase will cross the interface and evaporate into the CO_2 phase, if it is injected as dry CO_2. Similar to the CO_2 dissolution, the water vapor will eventually reach an equilibrium value in the vicinity of the fluid–fluid interfaces. So our first observation is that in regions of the porous medium where both fluids exist, we will necessarily have fluid–fluid interfaces across which mass can exchange, and the phases will reach equilibrium concentrations for the relevant components in each phase over these short length scales. We also note that dissolved CO_2 can reduce the pH of the brine, thereby possibly leading to various geochemical reactions. In addition, wet CO_2 is much more corrosive to metal surfaces (such as well casings) than dry CO_2. As such, component mass transfer can lead to a sequence of geochemical reactions that may need to be modeled.

The injection process will take place at relatively high pressures, so that near to the injection well the capillary pressure will be quite large. We know from considerations of capillary pressure and its relationship to saturation that at high capillary pressures, the system will tend to have wetting fluid close to, or at, its residual saturation, $s_b = s_b^{res}$. Therefore, we expect the region close to the well that is invaded by CO_2 to have brine at residual saturation. That brine will have dissolved CO_2 in it, with concentration that reaches equilibrium values relatively quickly. Similarly, the CO_2, with saturation $s_c = 1 - s_b^{res}$, will fill with water vapor, thereby creating "wet" CO_2 instead of the injected "dry" CO_2. In regions close to the well, dry CO_2 will continually displace the wet CO_2 and in so doing will evaporate some of the residual brine. Eventually, the water in the residual brine will all evaporate, and the only fluid remaining in the pore space will be dry CO_2 (along with some precipitated salts from the evaporated brine). This creates an expanding region of dry CO_2 in the vicinity of the injection well.

The injected CO_2 is driven into the formation by the pressure gradients generated by the pressure at the well, with its flow paths modified by capillarity and the associated relative permeability functions as well as heterogeneities in formation properties. The flow is also affected significantly by the strong density difference between the brine and CO_2, which leads to a buoyancy-driven upward movement of CO_2 that produces strong gravity segregation. This strong buoyant drive is one of the most important characteristics of the system because it gives a clear macroscopic spatial structure to the fluids. Buoyancy is also a major driving force in potential leakage scenarios, thereby highlighting the importance of the caprock formation above the injection formation. This buoyant segregation is a key element in approaches that allow simplifications to mathematical models for these systems, and we take full advantage of this in the models that we present in Chapters 4 and 5.

Once the injection ends, CO_2 is no longer being added to the system, and the fluids will begin to rearrange themselves in the formation. This includes locations where brine will flow back into CO_2-filled regions, which is an imbibition process. Just as residual brine saturation is produced during drainage (CO_2 displacing brine), residual CO_2 saturations are produced by the imbibition process. This leads to regions where (residual) CO_2 is trapped by capillary forces. This capillary trapping of CO_2, associated with the postinjection imbibition process, is important in the overall fate of the injected CO_2.

At this point, it is helpful to estimate the time required for the local-scale dissolution processes to take place as well as the vertical segregation of the fluid phases. The dissolution is driven by diffusion away from the actual fluid–fluid interface, and occurs over distances that might span a few pores, meaning a length scale on the order of a fraction of a millimeter to perhaps a few millimeters (say 10^{-3} meters). A simple diffusion calculation gives a time scale to reach close to equilibrium of hours to days, using an estimated diffusion coefficient of about 10^{-9} m^2/s. Therefore, for practical purposes, we can consider this local-scale mass transfer, within the region where both fluid phases exist, to be at equilibrium over the practical time scales associated with injection and transport, which are on the order of years to centuries. We can make similar arguments for pore-scale equilibration of capillary forces around fluid–fluid interfaces. We expect that on very small length scales, capillary equilibrium will occur quickly, and therefore, at these spatial scales, the process may be treated as always being at equilibrium.

We can perform a similar analysis for the buoyant segregation. If we consider the driving force in the fluid movement to be buoyancy, then a simple estimate of the upward velocity of buoyant CO_2 is given by $(k/\mu\phi)\Delta_\alpha \rho g$. With typical values for fluid properties taken from Section 3.2, with a porosity of about 0.1, and with a typical vertical travel distance of about 50 m, a formation with vertical permeability on the order of 10 milliDarcy would require about 1 year for buoyant-driven segregation. We note that relative permeability effects, especially at low brine saturations, can modify this time estimate significantly.

Once the CO_2 is vertically segregated by buoyancy, several additional factors may be considered. The first is whether or not the CO_2 will enter into the caprock. This becomes a simple question of whether or not the capillary pressure between the CO_2 at the top of the injection formation and the brine within the caprock exceeds the entry pressure for the caprock. If we imagine the CO_2 segregated at the top of the injection formation having a thickness $h_{T,I}$, and the brine surrounding that CO_2 plume being at hydrostatic pressure and also being connected to the brine in the caprock, then the capillary pressure along the caprock-formation boundary will be given by $\Delta_\alpha \rho g h_{T,I}$, where we have assumed the CO_2 is in vertical equilibrium and that the entry pressure in the aquifer is negligible compared to the entry pressure into the caprock. If we take an effective pore radius for invasion to be represented by r_{eff}, and we note that surface tension for CO_2–brine interfaces is about 0.035 N/m (see Section 3.4), then for a strongly wetting system we can use Equation (3.3) to estimate the entry pressure as $p_{entry}^{cap} \approx (0.07\,\mathrm{N/m})/r_{eff}$. Therefore, the entry pressure is exceeded when the CO_2 thickness exceeds $(0.07\,\mathrm{N/m})/r_{eff}\Delta_\alpha \rho g$. Shales, which often form competent caprock formation, have pore radii on the order of perhaps 0.1–0.01 μm. This leads to plume thicknesses on the order of at least several hundred meters to exceed capillary entry pressure. As such, we conclude that most competent caprock formations will prevent entry of CO_2 due to capillary exclusion.

The second factor to be considered is the representation of fluid saturations as a function of depth in the formation. That is, in moving from the bottom to the top of the formation, what values for phase saturations should we expect to observe? The image we have developed so far is that CO_2 is driven to the top of the formation by buoyancy, while the overall pressure drive from the injection operation continues to drive CO_2 radially outward from the well, displacing brine as it flows. At some distance from the injection well, gravity override will have driven the CO_2 upward so that at the bottom of the formation the pores are filled with brine, with $s_b = 1$. In the previous sections, we have constructed an image of the saturation profile near the groundwater table based on capillary pressure and vertical equilibrium, leading to the concept of a capillary fringe. The same concept can be applied to the CO_2–brine contact, as we will elaborate in Chapter 4. While models of capillary fringe profiles with the associated assumption of vertical pressure equilibrium are appealing, in part, because they are relatively easy to construct, the concept of capillary equilibrium on the larger scale deserves some comment. We have already argued that at the pore scale, we expect capillary equilibrium in the vicinity of an interface to be a reasonable approximation. However, we now ask under what conditions, or more specifically, after what period of time, will the capillary fringe achieve its full profile within the formation. This requires some additional assumptions about how the overall saturation profile evolves through the complex actions and interactions among pressure drive, buoyant drive, multiphase flow nonlinearities, and material heterogeneities. One way to estimate the time for a full capillary equilibrium to develop is to assume we can begin with a sharp interface, and that the capillary profile will develop from this initial condition. Such analysis provides a time scale for development of the profile. To do this, we can rely on highly resolved numerical simulations, which give estimates on the order of years to tens of years. Similar figures are obtained by choosing representative values so as to linearize the diffusion coefficient, and using a characteristic length for the capillary fringe of a few meters. These results indicate that the time for the segregated fluids to relax to a state of full capillary equilibrium can be significant, depending on the porous medium and length scales of interest.

As a final consideration, we return to the issue of component mass transfer. We have already argued that in regions where both fluid phases coexist, we expect concentrations of components in the phases to reach equilibrium values quickly. As such, we can model the interphase mass exchange as an equilibrium process. However, we can also have mass transfer along the large-scale interface associated with the transition from the region of full brine saturation to the CO_2 region, either with or without a capillary fringe. Along the lowermost extent of the CO_2, the brine will be saturated with dissolved CO_2. This dissolved CO_2 can diffuse away from that macroscopic interface, and migrate through the vertical thickness of the bottom region that is filled only with brine. While the local pore-scale length scale used earlier in this section was on the order of 1 mm, now the distance associated with diffusive transport is the thickness of the brine region, which might be on the order of 10–50 m. Now the analysis of time required for diffusion to reach equilibrium concentrations throughout this region becomes on the order of tens of thousands of years or more. Therefore, mass transfer across the column of brine-filled material cannot be modeled at equilibrium.

When CO_2 dissolves into brine, the density of the brine increases slightly. This larger density creates a layer of brine adjacent to the macroscopic interface that is denser than the brine below it. This leads to gravity instabilities in the system, which ultimately leads to enhanced mixing within the column of brine. Detailed simulations show how these instabilities develop and evolve. In general, the overall behavior of the dissolved CO_2 mixing begins like standard diffusion but eventually develops a fingered structure driven by gravity instabilities. During the initial diffusive period, the dissolved CO_2 spreads in proportion to $\sqrt{t}$, which is the case for the usual Fickian diffusion process. However, after the onset time for gravitational instability, fingered structures form, and the tips of these fingers of dissolved CO_2 move downward with a constant speed, and the CO_2 therefore spreads proportionally to t. This leads to the enhanced mixing for CO_2 and the associated reduction in time to reach a fully equilibrated brine region. This process of "convective mixing" can reduce the characteristic time to reach equilibrium by an order of magnitude or more. Note also that the increased density of the brine with dissolved CO_2 means that the CO_2 in the dissolved state no longer has a buoyant drive upward (like the separate-phase CO_2 has). Therefore, the risk of leakage is greatly reduced when CO_2 is in a dissolved state.

The processes of capillary trapping, dissolution ("solubility trapping"), and geochemical reactions (some of which may lead to precipitation of new carbonate rock—called "mineral trapping") are the mechanisms that become progressively more important after the cessation of injection of CO_2. These are shown schematically in Figure 3.17, which is taken from the 2005 special Intergovernmental Panel on Climate Change (IPCC) report on carbon capture and storage (CCS). That figure shows that during and immediately after injection, when most of the CO_2 exists as a buoyant separate phase, a competent caprock is required to keep the

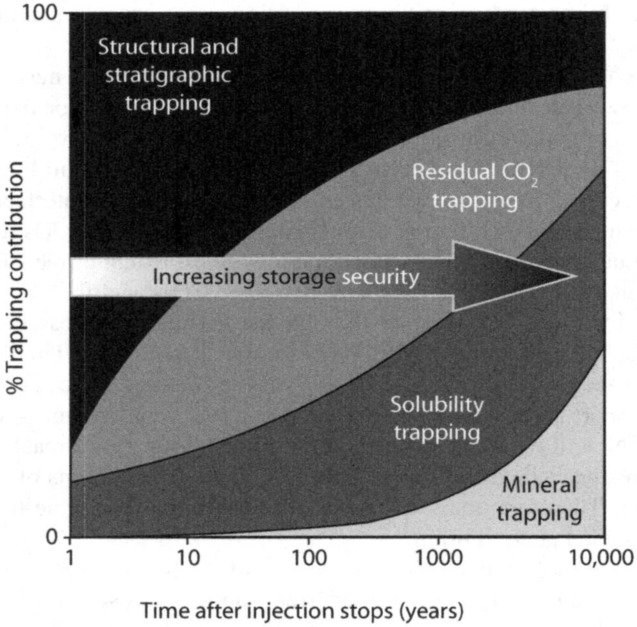

Figure 3.17 Schematic of the evolution of CO_2 from free phase to less leakage-prone states (from IPCC, 2005).

CO_2 in the injection formation. This is referred to as "structural and stratigraphic trapping." At longer time scales, more and more of the CO_2 becomes trapped as residual saturation, and progressively more dissolves into the aqueous phase. These processes greatly reduce the risk of leakage because the CO_2 changes from a lighter-than-water fluid driven upward by buoyancy to a fluid that is trapped and held in place by local capillary forces or is dissolved into the brine that then becomes more dense and is therefore pulled downward by gravity. If mineral trapping were to occur, this obviously produces a very stable form of trapped carbon. All of this means that storage security will increase with time after the cessation of injection.

We will return to Figure 3.17 in Chapter 5, where we will make a quantitative version of this figure for a specific storage location. We also consider various leakage scenarios associated with compromised stratigraphic trapping, for example, due to leakage along existing wells that penetrate the caprock, in Chapter 6. For now, the important message is that a number of physical and chemical processes can affect the CO_2, and it is important to understand the relative importance of each of them. This requires an understanding of, among other things, characteristic length and time scales for each process, so that decisions about how to model the processes can be made. We take up this idea in Chapter 4.

3.11 FURTHER READING

Multiphase Flow in Porous Media Textbooks and Chapters

DULLIEN, F.A.L. 1992. *Porous Media: Fluid Transport and Pore Structure*. San Diego, CA: Academic Press.

HELMIG, R. 1997. *Multiphase Flow and Transport Processes in the Subsurface*. Berlin: Springer-Verlag.

LAKE, L. 1996. *Enhanced Oil Recovery*. Englewood Cliffs, NJ: Prentice Hall.

MOREL-SEYTOUX, J. 1973. Two-phase flows in porous media. In: *Advances in Hydroscience*, Vol. 9, V.T. Chow, ed. New York: Academic Press, pp. 120–202.

Textbooks on Related Subjects

AZIZ, K.A. and A. SETTARI. 1979. *Petroleum Reservoir Simulation*. London: Applied Science Publishers.

CHAVENT, G. and J. JAFFRE.1986. *Mathematical Models and Finite Elements for Reservoir Simulation*. Amsterdam: Elsevier.

EWING, R.E. 1983. *The Mathematics of Reservoir Simulation*. Philadelphia: SIAM Publications.

FIROOZABADI, A. 1999. *Thermodynamics of Hydrocarbon Reservoirs*. New York: McGraw-Hill.

PINDER, G.F. and M.A. CELIA. 2006. *Subsurface Hydrology*. New York: Wiley.

Papers

BENNION, D.B. and S. BACHU. 2008. Drainage and imbibition relative permeability relationships for supercritical CO_2/brine and H_2S/brine systems in intergranular sandstone, carbonate, shale, and anhydrite rocks. *SPE Reservoir Evaluation & Engineering*, 11(3), 487–496.

BUCKLEY, S.E. and M.C. LEVERETT. 1942. Mechanism of fluid displacements in sands. *Transactions of the AIME*, 146, 107–116.

CHALBAUD, C., M. ROBIN, J.-M. LOMBARD, F. MARTIN, P. EGERMANN, and H. BERTIN. 2009. Interfacial tension measurements and wettability evaluation for geological CO2 storage. *Advances in Water Resources*, 32, 98–109.

FISCHER, U. and M.A. CELIA. 1999. Prediction of relative and absolute permeabilities for gas and water from soil water retention curves using a pore-scale network model. *Water Resources Research*, 35(4), 1089–1100.

PIRI, M. and M.J. BLUNT. 2005. Three-dimensional mixed-wet random pore-scale network modeling of two- and three-phase flow in porous media. *Physical Review E*, 71.

SPAN, R. and W. WAGNER. 1996. A new equation of state for carbon dioxide covering the fluid region from triple-point temperature to 1100 k at pressures up to 800 MPa. *Journal of Physical and Chemical Reference Data*, 25(6), 1509–1597.

SPYCHER N., K. PRUESS, and J. ENNIS-KING. 2003. CO_2-H_2O mixtures in the geological sequestration of CO2. I. Assessment and calculation of mutual solubilities from 12 to 100°C and up to 600 bar. *Geochimica et Cosmochimica Acta*, 67, 3015–3031.

Chapter 4

Large-Scale Models

Chapters 2 and 3 presented a number of fundamental concepts associated with fluid flow in porous media, ultimately producing a set of governing equations for different systems and, for some, solutions to those equations in the form of analytical expressions. The focus was on the mathematical representation of flow physics, and while CO_2–brine systems were used as examples where possible, the material was meant to apply to general porous media systems involving one (Chapter 2) or two (Chapter 3) fluid phases. These general models are often used to model CO_2 storage activities, with CO_2 and brine being two fluids that fit into a much more general modeling framework. In this chapter, we focus directly on mathematical modeling approaches for CO_2–brine systems, and more specifically on a particular approach to modeling that uses specific attributes of the CO_2–brine system to design models that are substantially simpler than the general models. In order to incorporate the important attributes of the CO_2–brine system, we will use a modeling framework that is referred to as the heterogeneous multiscale method (HMM). While somewhat more abstract than the approach in the previous chapters, it offers a consistent framework within which to consider the important length and time scales of the CO_2 storage problem, and to write formal mathematical relationships among variables defined over different scales. Within this framework, we develop sets of governing partial differential equations that reflect the multiscale nature of the problem, constrained by a set of simplifying assumptions that are particularly suited to the CO_2–brine system. Ultimately, some of these simplified equations can be solved analytically, while others require numerical solution. In all cases, our objective is to derive equations, and their solutions, at length and time scales that are relevant to practical questions about CO_2 storage systems.

As we have already seen in Chapter 3, the physics of CO_2 storage are described by a system of nonlinear partial differential equations. Those equations involve general mass balance equations augmented by material-dependent constitutive equations involving parameters like permeability, relative permeability, and compressibility as well as expressions for fluid properties like density and viscosity. The equations must be closed through specification of the domain on which the equations apply and by boundary conditions along the domain boundary. The result is a set of

Geological Storage of CO_2: Modeling Approaches for Large-Scale Simulation, First Edition.
Jan M. Nordbotten and Michael A. Celia.

nonlinear equations with parameters that vary in space over a wide range of length scales. The combined challenge of nonlinear differential equations together with heterogeneous parameters assures that exact solutions at fine scales of resolution are not possible.

Approximate solutions fall in two main categories. The first category involves simplification of the problem a priori, either by reduction of the number of processes (often referred to as *model reduction*), or by simplification of the domain and the equation parameters (generally referred to as *upscaling* or *multiscale modeling*). This approximation strategy, based on modification of the general governing equations, leads to a new set of governing equations which tends to be simpler to solve. However, the solution to this simpler problem will only be an approximation to the original problem, and the discrepancy between them may not be quantifiable.

The second category involves *numerical discretization*, where the differential equations are replaced by a finite number of discrete equations. These discrete equations can then be solved by a suitably powerful computer. This strategy is often more amenable to mathematical analysis, and frequently, results may be derived to prove that the approximation is accurate, given that a sufficient computational effort has been applied. Unfortunately, for practical problems relating to CO_2 storage, the computational effort needed for accurate solutions to be obtained will often exceed the capabilities of even the largest supercomputers. Computational solutions with very fine spatial resolution also suffer from a lack of data to define the parameters at these fine scales. As such, these models tend to be impractical for calculations over large spatial and temporal domains.

Of course, these two approaches are not mutually exclusive; for example, most reduced models will still need to be solved by numerical methods. Therefore, even though our approach will involve a range of model reductions, we will almost always see this in the context of subsequent numerical discretization. In this chapter, we focus on strategies for model reduction and the associated reduction of complexity for the governing equations. In Chapter 5, we consider solution methods for those equations.

4.1 SPATIAL AND TEMPORAL SCALES

In order to design an appropriate modeling strategy for a particular problem, it is important to consider the spatial and temporal scales involved, and how the physical processes and parameters of the system relate to these scales. Important scales include those associated with the domain, system parameters, and physical processes; and those associated with the solution to the problem. In many cases, the scales may not all be a priori quantifiable, and investigations such as laboratory experiments, or preliminary calculations, may be needed. While Section 3.9 began a discussion of length and time scales, we provide here a more complete discussion of scales, estimate the ranges of each important scale, and group them into reduced categories that are descriptive of the models we will develop. The ranges we present should be treated as estimates, and they are intended to be used only as general guidelines for the design of solution approaches. While there may be special cases where the estimates provided here are not appropriate, the values given should generally provide reasonable guidance for CO_2 injection and migration problems while giving important guidance for model development.

4.1.1 Spatial Scales

The largest spatial scale we will consider for geological storage of CO_2 is the length scale (greatest horizontal extent) of the aquifer system into which the injection occurs, which may be on the order of a hundred kilometers (10^5 m) or more. This scale is important, as diverse issues such as capacity estimates, liability, and computational domains all relate to this scale. At the other end of the spectrum, we consider the scale of the CO_2–brine interface in a single pore, which is on the order of micrometers (10^{-6} m). This scale governs processes involving capillary pressure, mass transfer, and relative permeability. The ratio of our largest to smallest scale is then about 10^{11}.

In Figure 4.1, we have indicated a number of the important length scales for the CO_2 storage problem. Most of these scales can be deduced from measurements. We identify the scale for fractures as the width of the fracture, which, like most properties, can be highly variable within a single fracture and among groups of fractures. We estimate fracture widths to range from less than 1 mm to several millimeters. By the length scale for wellbore flow, we mean the width of the flow paths within which fluid may flow along a wellbore, especially along a leaky wellbore. This length scale spans from fractures or cracks in the well cement along the outside

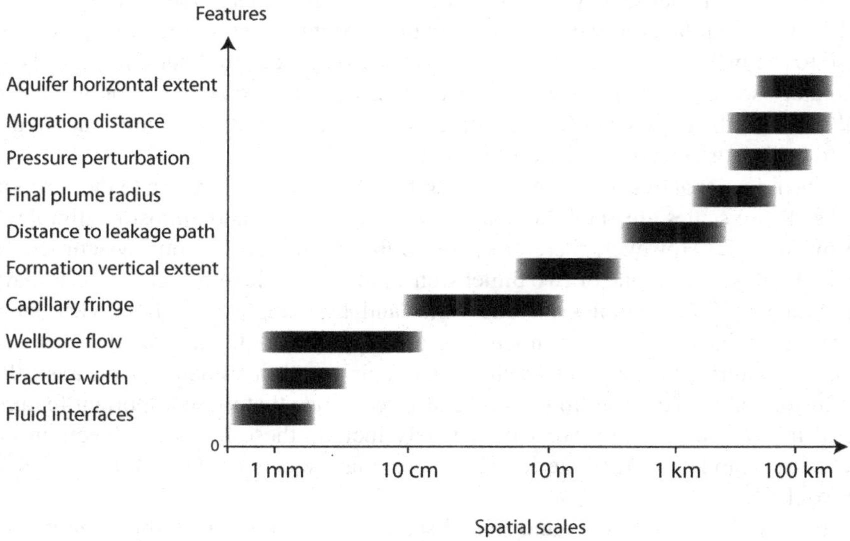

Figure 4.1 Estimates of spatial scales associated with CO_2 storage.

of the well casing, which have the length scales of fractures, all the way up to an open bore hole, which may have a radius of about 15–20 cm. As discussed in Chapter 3, the thickness of a capillary fringe region can be estimated from the capillary pressure curve. In the case of brine and CO_2, we expect this to range from a few centimeters to perhaps several tens of meters.

The vertical extent of the formation is self-explanatory, and is usually on the order of ten to several hundred meters. The distance to the nearest leakage feature refers to distances from the injection well to the nearest fault or fracture zone, or to the nearest leaky well. We estimate this, based largely on our studies of existing well fields in North America, to be between several hundred meters and several kilometers. For other parts of the world, where the legacy of oil and gas exploration and production is less intense, the distances may be somewhat larger, so we use as an upper bound 10–20 km. Finally, three of the largest length scales are defined as scales associated with the solution of the problem: the plume extent during injection, the extent of pressure perturbations associated with the injection, and the overall migration distance including postinjection migration. The ranges of values given here are based on analytical and numerical considerations (see Section 5.4), using values for hypothetical large-scale injection operations.

Based on the ranges of values indicated in Figure 4.1, we choose to identify four distinct spatial scales, which we term from largest to smallest: *macro*, *meso*, *micro*, and *nano*. The macro spatial scale is identified as above 100 m, and contains the three solution scales (migration, pressure perturbation, and plume scales). For practical purposes, we will require this scale to remain small enough to resolve the distance to nearest high-risk leakage path. The macroscale is the scale at which we wish to write governing equations and conduct numerical (or analytical) simulations.

The meso spatial scale is identified as spanning from tens to hundreds of meters and is strongly related to the vertical formation extent. Because the mesoscale is so close to the macroscale, we expect them to be strongly coupled, and the models we ultimately solve on the macroscale must therefore consider behavior on the mesoscale, even if the mesoscale is not resolved explicitly. We will use multiscale modeling to bridge the macro- and mesoscales.

The micro spatial scale covers the length scales from fractures up to the wellbore radius. These scales are small compared to the mesoscale, and almost negligible on the macroscale. However, there are some important objects in our system having small length scales in one or two dimensions, but a larger length scale in the remaining dimension(s). Examples include wells and fractures, which have very small openings but have larger vertical length scales, typically in the mesoscale range. We have characterized wells and fractures by their smallest dimension because that dimension tends to control flow. But we also recognize that these important features are of mixed length scale. We will typically include these kinds of objects in our macroscale models as discrete lower dimensional objects or through an upscaling approach.

Finally, we identify the nano spatial scale as the scale of CO_2–brine interfaces. Interfacial behavior at this scale determines microscale parameters such as capillary pressure, rates of mass transfer between phases, and relative permeability functions, which are important and ultimately need to be incorporated into the macroscale models. However, details of nanoscale processes usually do not have to be resolved explicitly, and the influence of this scale will only be seen through micro- and mesoscale parameterizations. Therefore, we do not represent the nanoscale explicitly in any of our macroscale models.

4.1.2 Temporal Scales

The largest temporal scale is defined not by a physical domain, but rather by the nonphysical factors like regulatory frameworks and operator demands. From an operator's point of view, the period of interest is the injection period, which is expected to be about 50 years. The regulatory guidelines are not clear yet, but their minimal temporal extent must be that of the injection operation, and there is likely to be some requirement regarding postinjection performance of the storage site. Therefore, processes that may not appear during the time scale of CO_2 injection may still be of interest for CO_2 storage and should be included in our consideration.

As discussed in Section 3.9, at the laboratory representative elementary volume (REV) spatial scale (micro spatial scale), both capillary equilibrium and compositional equilibrium occur over relatively short time scales. These are the two fastest (smallest) time scales identified in Figure 4.2. Wellbore leakage, or leakage along other preferential flow paths, tends to happen fairly quickly, and is listed as the next-fastest process of interest in the figure. Following again from the discussion in

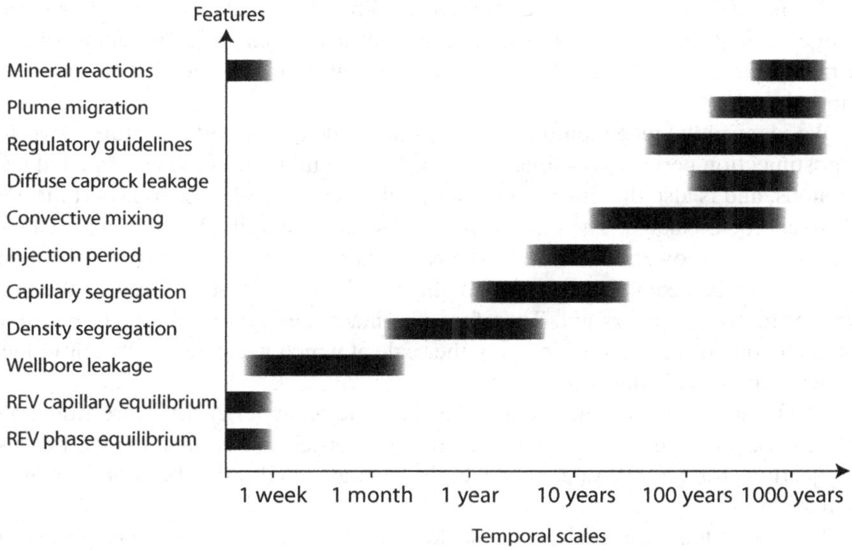

Figure 4.2 Estimates of temporal scales associated with CO_2 storage.

Section 3.9, density segregation is estimated to take on the order of years, while the formation of an equilibrium capillary fringe, at the large scale, is estimated broadly to be around 10 years in Figure 4.2. The injection period is self-explanatory, and the time scale associated with convective mixing, which includes both the onset of unstable mixing and the enhanced rate of mass transfer associated with the convection cells, is indicated to span times from tens to thousands of years. While the time scale of onset can be derived from stability analysis, the rate at which convection cells drive mass transfer can only be obtained through numerical calculations. Our estimates are based on calculations reported in the literature.

Pressure-driven leakage through caprock formations (referred to as diffuse leakage) is taken to be a slow process. For a water-wet caprock, if the CO_2 pressure does not exceed the entry pressure of the caprock, then capillary exclusion applies and no separate-phase CO_2 will flow through the caprock (except through possible preferential pathways like wells and faults). However, brine would leak slowly through a caprock with nonzero permeability, but the low permeability of these rocks makes the travel time through the caprock very long. This is indicated in the figure. Note that when we include diffuse caprock leakage at this time scale, we are thinking of transport times. Caprock leakage may be important at much earlier times due to its impact on the pressure field. Finally, the longest time scales are the time scales for regulatory guidelines and for long-term (postinjection) plume migration. Mineral reactions deserve special mention, in appearing both at long and short time scales. This reflects fast reactions with calcite, as well as slow reactions that ultimately form carbonate rocks. For the application of CO_2 storage in porous media, it is generally accepted that there are few mineral reactions happening at intermediate time scales.

Based on the ranges of values in Figure 4.2, we identify five temporal scales, with the term *mega* designating the largest scale. These temporal scales do not necessarily have a one-to-one correspondence with the spatial scales—for example, leakage through a wellbore, which covers a vertical extent of a formation (within the meso spatial scale), may be very rapid, and will be associated with the micro temporal scale.

We identify the mega temporal scale as the scale above half a century, covering the postinjection period. This time scale coincides with the time scale of global CO_2 emissions, and is also the time scale of mineral reactions and mass transport through diffuse caprock leakage. The macro temporal scale is identified as the time scale of injection, with a lower end at about 1 year. It is at this time scale that operational decisions will be taken, and it is also the time scale likely to be covered by the most stringent regulatory oversight. Therefore, the macro temporal scale takes a similar role to the macro spatial scale in being the scale at which we aim to write our models and perform our calculations.

The meso temporal scale is defined as the time scale of density segregation, and will be dependent on not only the formation properties and vertical extent, but also the injection strategy and rate. However, in general, it will often be shorter than the macroscale.

The micro temporal scale is the scale of wellbore and fracture leakage. Again, this scale is highly problem-dependent, but for high-risk leakage pathways, it will in general be shorter than the mesoscale. It is of particular interest from the perspective of monitoring and evaluation of risk during the macroscale analysis.

Finally, we identify the nano temporal scale as the time scale of equilibration of processes within an REV, which actually corresponds to the micro spatial scale. These processes will be considered to be in equilibrium at all larger temporal scales.

4.1.3 Implication of Spatial and Temporal Scales

The importance of identifying spatial and temporal scales lies in the guidance they give for modeling. As we have emphasized, our primary goal is to construct models at the macroscale (both temporal and spatial). Our analysis of scales then guides us in our application of model reduction, multiscale modeling, and application of numerical approximations.

From the temporal scales of interest, we see that mass transport associated with diffuse caprock leakage and mineral reactions are not expected to be of importance in a macroscale model. We will simplify our models by neglecting these processes (although, as already mentioned above, the diffuse leakage of brine through caprock can impact the pressure solutions on the macro time scale, as we considered in Section 2.4 in Cases 3 and 4). Further, the nanoscale is identified as too small to resolve, and is also expected to equilibrate quickly relative to the macroscale. As such, we will always assume that nanoscale processes are in an equilibrium state.

While not fully resolved, we will capture the micro and meso spatial scales using lower dimensional and multiscale representations, respectively. Our multiscale approach to capture the mesoscale will be described in the next section, and we consider representation of wells in Chapters 5 and 6.

In the models presented herein, we will mostly consider applications where the micro and meso temporal scales are small relative to the macroscale, and therefore assume that processes at the micro and meso temporal scales are in equilibrium, just as we have already assumed this at the nanoscale.

Megascale models are not the primary emphasis of this book, but we keep this scale in mind in Section 5.4.2 where a numerical study of late time model predictions is conducted. Similarly, mesoscale descriptions may be relevant for some problems, such as at early injection stages or for small aquifers.

4.2 MULTISCALE MODELING

Based on the analysis of spatial and temporal scales, and on our interest to develop models that are able to address practical issues associated with predictions over large spatial and temporal scales, we have chosen to develop mathematical models for the CO_2 storage problem at the macro spatial scale and the macro (and sometimes mega) temporal scale. It is these scales we have in mind when we refer to large-scale simulation. Because the macro spatial scale is defined to be on the same order as, or larger than, the vertical extent of an aquifer, the macro model will not resolve the vertical length scale within an aquifer. As such, we look to vertically integrated models, where the mesoscale behavior along the vertical dimension of a formation is incorporated into the macroscale equations via multiscale concepts. Indeed, the main challenge in deriving a macroscale model for CO_2 storage lies in the effective representation of the meso and micro spatial scales. Our macroscale model will describe the horizontal spread of CO_2 and brine, within a given aquifer, and as such will be a model defined over two spatial dimensions corresponding to the areal dimensions of a given aquifer. When multi-aquifer systems are considered, we will resolve the layered structure within the sedimentary sequence and thereby combine a series of two-dimensional aquifer models into a macroscale three-dimensional model which is sometimes referred to as "quasi-three-dimensional."

Modeling the same physical phenomena at different scales is a classical problem in mathematical representations of physical systems. A relevant example involving flow in porous media is the description of flow at the nano spatial scale versus the description of the same flow at the micro or meso spatial scales. At the nanoscale (which for us corresponds to the pore scale), equations of fluid motion, such as the Navier–Stokes equations, would be written, and the individual grains of the solid material would define boundaries for the tortuous, pore-scale domain of flow. At the micro- or mesoscale, individual grain boundaries are no longer represented explicitly. Rather, parameters like porosity and permeability arise, and the equation is transformed from Navier–Stokes to Darcy's law. Both of these models can describe the flow process well, but the equations and the parameters of the equations are different. The fact that these two models successfully describe the same flow implies a relationship between models. If we define $\boldsymbol{u}$ to be the flow velocity from the Navier–Stokes equations, and $\boldsymbol{U}$ to be the Darcy flux, we expect some relationship between the finer-scale ($\boldsymbol{u}$) and the coarser-scale ($\boldsymbol{U}$) variables such that $\boldsymbol{U} = \mathcal{C}\boldsymbol{u}$, where (as in Chapters 2 and 3) $\mathcal{C}$ denotes an operator that compresses detailed information from a finer scale to a bulk or average measure at the coarser scale. Typically, such a compression operator is defined by either integration (averaging) of the fine-scale function or by finite sampling of specific values of the fine-scale function.

Unlike the case with the Darcy equation and the Navier–Stokes equations, the equations that model the coarse-scale are often not known a priori, and our objective is to derive them. During the last decade, the mathematics community has found a renewed interest in the problem of simulating on a coarse scale, even if the coarse scale equations cannot be explicitly derived. One of the approaches that follow this direction was introduced by Weinan E and Björn Engquist as the so-called heterogeneous multiscale method, or HMM. In this chapter, we will use the language of HMM to describe scale interactions. While these ideas were introduced in the previous chapters, we now exploit the multiscale framework to a much greater extent. This will allow us to generalize our integrated models.

The main goal of HMM is to provide a framework where fine-scale models can be used to inform coarse-scale simulation. In other words, we presume that we have some incomplete coarse-scale model, such as a conservation law, but we are missing some of the information at the coarse scale, such as the proper functional form for the flux expression. The HMM framework provides an implicit description of the coarse flux expression based on a fine-scale model. The main ingredients in this approach are a compression operator $\mathcal{C}$, a fine-scale model, an incomplete coarse-scale model, and a reconstruction operator $\mathcal{R}$. The reconstruction operator takes a coarse-scale solution, and reconstructs some approximation of a fine-scale solution from which the fine-scale model can base its calculation to inform the coarse-scale calculation. The reconstruction and compression should be consistent, so that if we first reconstruct and then compress (that is to say, apply the combined operator $\mathcal{CR}$), it leaves the coarse function unchanged. We refer to this as the *consistency requirement* for the HMM approach. This requirement will be enforced as we proceed through the following derivations. Note that, in general, the converse of the consistency requirement does not hold because the reconstruction may not be unique.

To illustrate the HMM methodology, we consider the unconfined aquifer problem presented in Chapter 2, which was further discussed in Section 3.5. Our fine-scale model is defined to correspond to the laboratory scale, taken to be a three-dimensional domain where Darcy's law is valid, and the coarse scale is taken to be the lateral extent, defined by a two-dimensional (horizontal) domain for which we wish to derive an equation that does not include any explicit representation of the vertical dimension. This is illustrated in Figure 4.3, where we have emphasized that the compression operator takes us from three-dimensional governing equations to two-dimensional equations. We also see that the reconstruction and compression operators in general do not return the actual fine-scale description, unless the fine-scale description is exactly a reconstructed expression.

We let the impermeable bedrock be horizontal, and located at $z = 0$. Our coarse-scale variable is then the effective fluid content of the groundwater mound (taken as the height to the water table), denoted as a height $\zeta = \zeta(x_1, x_2)$. We define compression operators for the water saturation, s_w, whose purpose is to take information at the fine scale and integrate it to a coarse-scale measure. In this case, the compression involves vertical integration, which will lead to a two-dimensional (x_1, x_2) system. To distinguish among different kinds of compression operators, we will use a subscript to identify the coarse-scale variable that results from the compression operation. The variable to which the compression operator is applied is shown explicitly in the operation. For saturation, the coarse-scale quantity produced by the

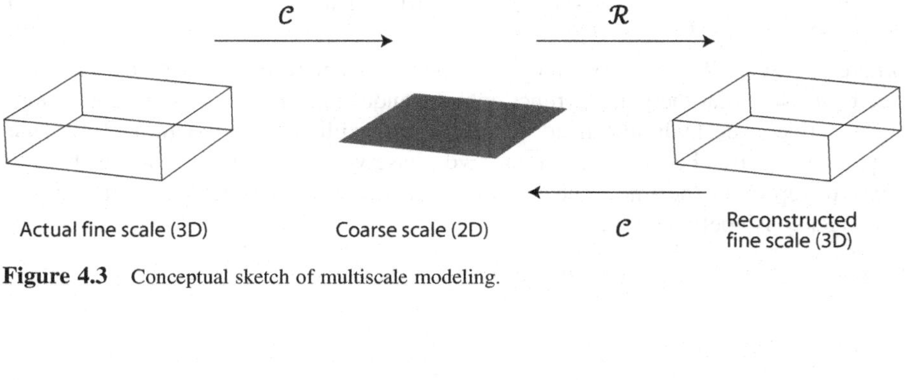

Figure 4.3 Conceptual sketch of multiscale modeling.

compression operation is the height of the mound, ζ. Thus, we have the following definition of C_ζ:

$$\zeta = C_\zeta s_w \equiv \int_0^{\zeta_T} s_w dz.$$

Here ζ_T denotes the top of the domain, taken to correspond to the land surface. We will not concern ourselves with heterogeneity of porosity in this section, and let ϕ be constant. We also note that, for this problem, the integrated fine-scale saturation, which produces a kind of coarse-scale saturation, is conceptually equal to the saturated thickness (the groundwater thickness) when water in the unsaturated zone is taken to be negligible. For a horizontal bottom boundary, the integrated saturation is simply equal to the elevation of the top of the mound above the bottom boundary.

Once we have defined compression operators and associated coarse-scale variable, we can write the general mass balance equation that we know holds at all scales. We write it here as a volume balance, assuming the system is incompressible, the porosity is constant in the vertical direction, and that there is no flux through the top or bottom:

$$\phi \frac{\partial \zeta}{\partial t} + \nabla_{\|} \cdot \boldsymbol{U} = 0. \tag{4.1}$$

This is an incomplete equation in that we need to specify an explicit functional dependence for the coarse-scale flux term before the equation can be solved. The definition of the functional dependence of the coarse-scale flux function should be written as a function of only coarse-scale variables, and it should be constrained to give consistent (mass conservative) flux values. For this example, we choose to write the coarse-scale flux vector, $\boldsymbol{U}$, as a function of the coarse-scale fluid content, that is, $\boldsymbol{U}(\zeta)$. This choice implies that we make the tacit assumption that the variables that define the fine-scale flux $\boldsymbol{u}$, which in the present case are h and s_w, can be adequately approximated from the coarse variable.

Using the fact that the coarse variable ζ is defined from the fine scale via the operator C_ζ, a compression operator for the fine-scale volumetric flux vector $\boldsymbol{u}$, denoted by C_U, can be uniquely determined such that Equation (4.1) is consistent with the fine-scale volume conservation equations from Chapter 3. It then follows that,

$$\boldsymbol{U} = C_U \boldsymbol{u} \equiv \int_0^{\zeta_T} \boldsymbol{e}_{\|} \cdot \boldsymbol{u}(s_w, h) dz, \tag{4.2}$$

where $\boldsymbol{e}_{\|} = [\boldsymbol{e}_1 \ \boldsymbol{e}_2]^T$ serves to remove the z-component from the three-dimensional vector $\boldsymbol{u}$. Note that the general functional dependence of $\boldsymbol{u}$ on fine-scale saturation, s_w, and fine-scale hydraulic head, h, is denoted explicitly in Equation (4.2). That dependency arises because we use the two-phase version of Darcy's law at the fine scale to represent the fine-scale fluxes. It is written here (assuming isotropic and homogeneous coefficients) as,

$$\boldsymbol{u}(s_w, h) = -\lambda(s_w)\kappa \nabla h.$$

Note that the coarse-scale flux could properly be referred to as an "integrated flux vector." We emphasize that the compression of the flux vector cannot be defined a priori, but must be chosen such that it is consistent with the compression operator for the coarse variables and the conservation equations we impose at the coarse and fine scales.

Given this functional dependence, we can define the relationship $\boldsymbol{U}(\zeta)$ by replacing the dependence on the actual fine-scale solution in Equation (4.2) with a dependence on reconstructed variables. The functional form of the relationship $\boldsymbol{U}(\zeta)$ will then depend on the specific reconstruction operators chosen. In this case, reconstruction of the fine-scale saturation would be achieved by the operator $\mathcal{R}_s$, and reconstruction of the fine-scale hydraulic head would be accomplished by the operator $\mathcal{R}_h$. Note that, in terms of notation, the reconstruction operators follow the same convention as the compression operators, in that the subscript will denote the fine-scale variable that is being reconstructed, and the coarse-scale variable to which the operator is applied will be shown explicitly when writing the operation. Once these operators are defined, we can write $\boldsymbol{U}(\zeta)$ formally as

$$\boldsymbol{U}(\zeta) = \int_0^{\zeta_T} \boldsymbol{e}_{\parallel} \cdot \boldsymbol{u}(\mathcal{R}_s\zeta, \mathcal{R}_h\zeta)\, dz. \tag{4.3}$$

For this simple problem of a groundwater mound, a reasonable modeling choice for the reconstruction operators may be the second operator discussed in Section 3.9. This yields a constant saturation profile in the saturated region along with a constant hydraulic head. The suggested definitions are therefore,

$$\widehat{s_w} = \mathcal{R}_s^2\zeta \equiv \begin{cases} 1 & \text{if} \quad z \le \zeta \\ 0 & \text{if} \quad z > \zeta \end{cases}$$

$$\hat{h} = \mathcal{R}_h\zeta \equiv \begin{cases} \zeta & \text{if} \quad z \le \zeta \\ 0 & \text{if} \quad z > \zeta \end{cases}$$

The superscript 2 on the saturation reconstruction denotes the second reconstruction choice from Section 3.9. We have used a hat on the fine-scale variables to denote that they are reconstructed, and thus, in general, they differ from the actual fine-scale variables (denoted without hat). With this choice of reconstruction operators, we see that we can evaluate Equation (4.3) explicitly as

$$\boldsymbol{U}(\zeta) = -\kappa\lambda(1)\zeta\nabla_{\parallel}\zeta = -\frac{\lambda(1)\kappa}{2}\nabla_{\parallel}\zeta^2. \tag{4.4}$$

Equations (4.1) and (4.4) are the coarse-scale equations resulting from the HMM approach, with the particular choice of reconstruction operators. These equations are equivalent to the equations for an unconfined aquifer given in Chapter 2. Note that we could have obtained the same result if we had defined the compression operator as sampling the fine-scale head at the bottom of the formation, $\zeta = \mathcal{C}_{\zeta}h \equiv h(z = 0)$, instead of the vertical integration. However, without the interpretation of ζ as a conserved quantity, we would not have had the justification for Equation (4.1), the coarse-scale equation of our model. Choice of compression operators should recognize that balance laws will always be written at the coarse scale, and the coarse-scale variables will need to be related to those balance laws. It is also worth pointing out that a good choice for a reconstruction operator is one that produces reconstructed fine-scale variables that are close to the true fine-scale variables while requiring relatively little computational effort to reconstruct from the compressed variables.

For this simple example, it is instructive to consider the other reconstructions from Section 3.9. The topic of capillary fringe reconstruction, the third reconstruction, will be discussed in detail later; therefore, we now only point out that if we choose the first reconstruction for saturation, $\mathcal{R}_s^1$, the flux vector takes the form

$$U(\zeta) = -h_{T,B}\,\lambda(\zeta / h_{T,B})\kappa \nabla_{\|}\zeta,$$

where $h_{T,B}$ is the thickness between the top (T) to the bottom (B) boundary of the vertical domain. The flux arising from reconstruction $\mathcal{R}_s^1$ therefore not only results in quantitative differences, but for nonlinear mobility functions there is also a qualitative difference in the nonlinear dependence on ζ. This points clearly to the need to incorporate as much information as possible in the vertical reconstruction operation.

We close this section by summarizing our notation. We will continue the convention from the previous sections, and consistently use lowercase and uppercase letters to denote the fine and coarse scales, respectively. However, we will see in Chapter 5 that what constitutes the fine scale, and what constitutes the coarse scale, may be context-dependent. In addition, we will use a hat to indicate a reconstructed fine-scale variable to distinguish it from the true fine-scale variable. Finally, we note that the fact that $\mathcal{C}$ and $\mathcal{R}$ are only one-sided inverses may be expressed as $\widehat{s_w} = \mathcal{R}_s\zeta = \mathcal{R}_s\mathcal{C}_\zeta s_w \neq s_w$, while, however, by applying $\mathcal{C}_\zeta$ to the previous statement, we see that $\zeta \equiv \mathcal{C}_\zeta s_w = \mathcal{C}_\zeta \widehat{s_w} = \mathcal{C}_\zeta \mathcal{R}_s\zeta = \zeta$. Because this latter identity applies to all coarse-scale variables, we can write it more generally as $\Xi = \mathcal{C}_\Xi \hat{\boldsymbol{\xi}} = \mathcal{C}_\Xi \mathcal{R}_\xi \Xi$, where Ξ and $\boldsymbol{\xi}$ are the vectors containing all coarse- and fine-scale solution variables, respectively.

4.3 COARSE DOMAIN, VARIABLES, AND CONSERVATION EQUATIONS

We are now ready to define a set of coarse models for CO_2 storage. Based on our analysis of scales, we will define these models at the macro spatial scales wherein the vertical thickness of a formation is below the model scale. Therefore, we will define our coarse models using coarse variables derived from compression across the vertical dimension. Our choice of coarse-scale variables will be associated with a set of conservation equations; however, as we have seen, these alone will not be sufficient to give a closed system of equations. The system can only be closed by relating coarse-scale variables to one another, for example, the coarse-scale flux term to coarse-scale measures of potential with other coarse-scale parameters (coarse permeability, for example) also involved. This will be achieved through definition of reconstruction operators, which are meant to provide physically reasonable representations at the fine scale. These reconstruction operators allow fine-scale fluxes to be calculated, and these reconstructed fine-scale fluxes are then compressed back to the coarse scale to provide a closed system of equations that can be solved. Indeed, we will see that for most models, the coarse-scale variables will always be obtained from the fine scale by essentially the same compression operator. The determination of reconstruction operators constitutes the only difference between the models, and is where all understanding of subscale processes is embedded. The continuous evolution in complexity of reconstruction operators constrained by compression to coarse-scale variables is the essence of our approach.

The general outlines of the problem at hand, together with some of the notation we will exploit, can be seen in Figure 4.4. Note that we are now allowing somewhat more general geometries in the problem; in particular, the formation of interest can be tilted away from the horizontal. Our coordinate axes will in general align with this tilt, and we make the distinction between the coordinate axis, x_3, over which we will perform our compression (often integration) and the vertical coordinate, z (see Figure 4.4). Note that the top (T) and bottom (B) boundaries of the formation, denoted by ζ_T and ζ_B, are defined with respect to the x_3-coordinate, not the z-coordinate. For many geological formations, as was noted in Chapter 1, the difference between the x_3- and z-directions will be at most a few degrees. We will also introduce other surfaces that can vary with x_1, x_2, and possibly t. Those surfaces, like ζ_T and ζ_B, will always be denoted by the Greek letter ζ with an appropriate subscript.

Our fine-scale model for two-phase, two-component flow of CO_2 and brine, as given in Chapter 3, is taken to be valid for the micro and meso spatial scales. The primary variables in this fine-scale model are the component mass densities m^i, from which the secondary variables (mass fractions m^i_α; saturations s_α; densities ρ_α;

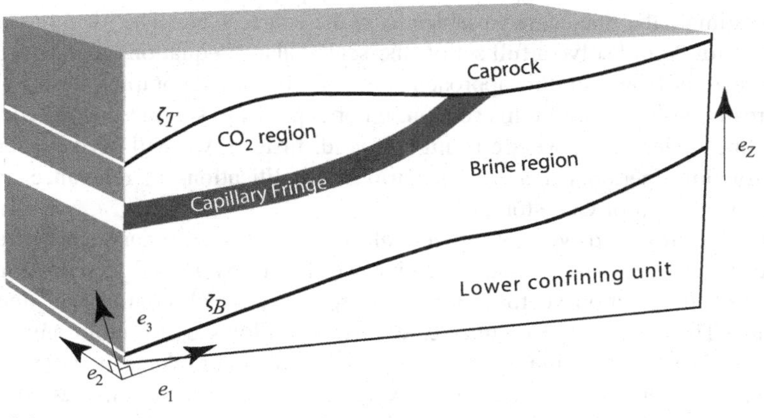

Figure 4.4 Schematic of the CO_2–brine system, with coordinate systems and pressure datum.

porosity ϕ; and pressures p_α) can be calculated. For the coarse scale, we choose a similar primary variable and define the coarse mass densities M^i through the compression operator for component mass, $\mathcal{C}_{M^i}$:

$$M^i = \mathcal{C}_{M^i} m^i \equiv \sum_\alpha \int_{\zeta_B}^{\zeta_T} \rho_\alpha \phi s_\alpha m_\alpha^i dx_3 = \int_{\zeta_B}^{\zeta_T} m^i dx_3. \tag{4.5}$$

Notice that, while m^i can be interpreted as the mass of component i per volume of total porous medium, M^i represents the mass of component i in the entire interval $\zeta_B \leq x_3 \leq \zeta_T$, per unit area in (x_1, x_2). Because mass is conserved, M^i satisfies a conservation law at the coarse scale, which we express as

$$\frac{\partial M^i}{\partial t} + \nabla_\| \cdot \boldsymbol{F}^i = \Psi_\Sigma^i. \tag{4.6}$$

Here, $\boldsymbol{F}^i$ is a coarse mass flux and Ψ_Σ^i is a coarse source or sink term. We require that each term in Equation (4.6) is consistent with the corresponding fine-scale term, and thus we have

$$\boldsymbol{F}^i = \boldsymbol{e}_\| \cdot \sum_\alpha \int_{\zeta_B}^{\zeta_T} \rho_\alpha \boldsymbol{u}_\alpha m_\alpha^i dx_3 \tag{4.7}$$

and

$$\Psi_\Sigma^i = \int_{\zeta_B}^{\zeta_T} \psi^i dx_3 + \Psi_B^i - \Psi_T^i. \tag{4.8}$$

We recall from Chapter 3 that the fine-scale phase fluxes can be expressed through the two-phase extension of Darcy's law,

$$\boldsymbol{u}_\alpha = -\boldsymbol{\lambda}_\alpha(s_\alpha)\boldsymbol{k}(\nabla p_\alpha - \rho_\alpha \boldsymbol{g}). \tag{4.9}$$

We see immediately that the preceding equations do not form a closed coarse-scale model because the coarse-scale flux is dependent on the actual fine-scale variables, which are not known. Our approximation is as expected to use reconstruction operators to estimate the fine-scale variables from the coarse scale. The ideal reconstruction operator would solve a full set of fine-scale balance equations to determine the reconstruction. However, this approach is too nonlinear and complex for the closed-form treatment we desire in this section; it corresponds to solving the full equations on the fine scale, which we are trying to avoid. Instead, we will derive a series of reconstruction operators that exploit various simplifications of relevance to macroscale simulation of CO_2 storage.

Because flow is driven by a potential law, a useful secondary coarse variable will be the coarse phase pressures, which we denote by $P_\alpha = \mathcal{C}_{P_\alpha}\boldsymbol{\xi}$. Here we have again used the solution vector $\boldsymbol{\xi} = \{m^i, p_\alpha, s_\alpha, \ldots\}$ which contains all fine-scale variables. This general set of fine-scale variables allows us to avoid any a priori restrictions on construction of the compression operator. We can define similar compression operators for coarse-scale variables like porosity, densities, and mass fractions, which are denoted by $\Phi = \mathcal{C}_\Phi \boldsymbol{\xi}$, $\varrho = \mathcal{C}_\varrho \boldsymbol{\xi}$, and $M_\alpha^i = \mathcal{C}_{M_\alpha^i}\boldsymbol{\xi}$, respectively.

We note that if we can define a reconstruction from the coarse primary variables alone, such that $\hat{\xi} = \mathcal{R}_{\xi}M$, then from the compression operator we could obtain constitutive functions on the coarse scale, such as a relationship between coarse phase pressures and coarse masses, for example, $P_\alpha(M) = \mathcal{C}_{P_\alpha}\mathcal{R}_{\xi}M$. This is indeed just a special case of the consistency requirement $\Xi = \mathcal{C}_{\Xi}\mathcal{R}_{\xi}\Xi$. However, such constructions are not always possible; in particular, pressure cannot be reconstructed from masses when the compressibility is neglected. In such cases, we may be motivated to choose pressure as an independent variable. Even when fluids are compressible, coarse pressure usually plays an important role in coarse-scale equations. In much of the remainder of this chapter, we will consider specific compression and, especially, reconstruction operators, and see how they fit into coarse-scale models for the CO_2 storage problem.

We conclude this section by making some comments on the adjectives we use to describe variables. We will consistently use the expression *primary* to denote coarse-scale variables that represent conserved quantities. In this chapter, we only consider mass and fluid transport; therefore, primary variables will be component masses. If we extended our models to include energy transport, we would also have to work with energy (or enthalpy) as a primary variable as discussed in Section 2.2.3. *Secondary* and *tertiary* variables refer to all other variables. For clarity, we distinguish variables that are meaningful for all reconstructions (such as pressure and saturation) as secondary, while those specific to a particular reconstruction (such as various sharp interfaces, see Section 4.6) are referred to as tertiary. Introduction of secondary and tertiary variables always necessitates additional modeling assumptions. An example of a modeling assumption associated with the secondary variable pressure is vertical equilibrium, while a model using tertiary variables such as sharp interfaces may involve a hysteresis model, as in Section 4.6.

Contrasting the use of primary, secondary, and tertiary, which emphasizes the role of a variable with respect to modeling, we have the concept of independent and dependent variables. We use these adjectives to identify the variables for which the equations are solved (independent), and those variables which are obtained from constitutive relations (dependent). We will see that for our models, secondary and tertiary variables are often chosen as the independent variables, demonstrating that the final form of a coarse-scale model does not have to be written in terms of primary variables.

4.4 DUPUIT RECONSTRUCTION OF PRESSURE

As mentioned in the previous section, fluid pressure almost always plays an important role in fluid flow models, and it is often convenient to consider pressure as an independent, secondary variable on the coarse scale. This will be the case for several of our models discussed below, and we therefore introduce the most important assumption we will make with respect to pressure, namely the Dupuit assumption. A special case of the Dupuit assumption, the vertical equilibrium assumption, has already been applied extensively in Chapters 2 and 3. The Dupuit assumption assigns a pressure field such that flows perpendicular to the aquifer boundaries are ignored. For the system shown in Figure 4.4, this means that flows in the x_3-direction are neglected. We denote the reconstruction operators that are based on the Dupuit assumption with a superscript D. We will not use superscripts on compression operators, since in general, we will only consider a single compression for each coarse variable.

Spatial averaging of pressure is difficult to justify physically, in part because pressure is not an additive quantity (like mass or volume). Furthermore, under equilibrium conditions, pressure is not a constant function of elevation. Therefore, we choose the coarsening (compression) operator for pressure based on subsampling instead of integration, and we define the sampling as the pressure at a datum $x_3 = \zeta_P$:

$$P_\alpha = \mathcal{C}_{P_\alpha} p_\alpha \equiv p_\alpha(x_3 = \zeta_P).$$

Note that care must be taken to ensure that either the datum is within the domain of definition for the fine-scale pressure p_α, or that a suitable extension of the pressure outside of the domain of definition is given.

The vertical equilibrium assumption is essentially an assumption of negligible flow in the vertical direction. In our setting, it is more natural to think of negligible flow perpendicular to the formation, which we denote as the Dupuit assumption. Thus,

$$0 \approx \boldsymbol{u}_\alpha \cdot \boldsymbol{e}_3 = -\lambda_{\alpha,3}(s_\alpha) k_3 \left(\frac{\partial p_\alpha}{\partial x_3} - \rho_\alpha \boldsymbol{g} \cdot \boldsymbol{e}_3 \right). \tag{4.10}$$

Here, and in the rest of this chapter, we assume that x_3 is aligned with the direction of one of the principle components of both permeability and mobility:

$$\boldsymbol{k} = \begin{bmatrix} \boldsymbol{k}_{\|} & \boldsymbol{0} \\ \boldsymbol{0} & k_3 \end{bmatrix} \quad \text{and} \quad \boldsymbol{\lambda}_\alpha = \begin{bmatrix} \boldsymbol{\lambda}_{\alpha,\|} & \boldsymbol{0} \\ \boldsymbol{0} & \lambda_{\alpha,3} \end{bmatrix}.$$

Equation (4.10) is satisfied if the dependence of pressure on x_3 is defined by integration of the expression in the parenthesis in the above equation. If the integration begins at the pressure datum surface, and we assume that the approximation can be replaced by equality, then the appropriate expression is

$$0 = \int_{\zeta_P}^{x_3} \left(\frac{\partial p_\alpha}{\partial x_3'} - \rho_\alpha \boldsymbol{g} \cdot \boldsymbol{e}_3 \right) dx_3'.$$

This may be written equivalently as

$$p_\alpha(x_3) - p_\alpha(\zeta_P) = (\boldsymbol{g} \cdot \boldsymbol{e}_3) \int_{\zeta_P}^{x_3} \rho_\alpha dx_3'.$$

With the replacement of the approximate equality in Equation (4.10) with the strict equality, we have a well-defined reconstruction on the fine scale which can serve as our reconstruction operator. That is, noting that the definition of the coarse-scale pressure is the sampled pressure at the datum, $P_\alpha = p_\alpha(\zeta_p)$, we have the Dupuit reconstruction operator for fine-scale pressure,

$$\widehat{p_\alpha} = \mathcal{R}^D_{p_\alpha} P_\alpha \equiv P_\alpha + (\boldsymbol{g} \cdot \boldsymbol{e}_3) \int_{\zeta_P}^{x_3} \rho_\alpha dx_3'. \tag{4.11}$$

When the density is constant in the vertical direction, the reconstructed pressure $\widehat{p_\alpha}$ becomes a linear function in x_3.

4.5 MODELS FOR IMMISCIBLE, INCOMPRESSIBLE PROBLEMS

This section will describe vertically integrated models for immiscible fluids with constant density, where the solid matrix is incompressible. Our presentation in this section will be fairly general, thereby allowing us to present both simplifications and extensions of the model in the following sections.

We will denote choices of compression and reconstruction operators that are particular to immiscible and incompressible problems by a superscript *II*.

The assumption of immiscibility implies that the fine-scale solution satisfies $m^i_\alpha = m^{i,0}_\alpha$, where the superscript "0" denotes the initial values of the mass fractions. Further, constant density of the fluid implies that $\rho_\alpha = \rho^0_\alpha$ for some constants ρ^0_α. It is natural to construct compression operators such that these properties are conserved on the coarse scale, thus $M^i_\alpha = \mathcal{C}^{II}_{M^i_\alpha} \xi \equiv m^{i,0}_\alpha$; and $\varrho_\alpha = \mathcal{C}^{II}_{\varrho_\alpha} \xi \equiv \rho^0_\alpha$. The corresponding reconstruction operators simply return the constant values to the fine scale, for example, $\widehat{m^i_\alpha} = \mathcal{R}^{II}_{m^i_\alpha} M^i_\alpha \equiv M^i_\alpha$.

The coarse-scale porosity is defined in all our models using integration as

$$\Phi = \mathcal{C}_\Phi \xi \equiv \int_{\zeta_B}^{\zeta_T} \phi dx_3.$$

Note that we have not scaled this equation by the integration length, thus the coarse-scale porosity is bounded not by 1, but by $\Phi \leq h_{T,B}$, and has units of length. We will sometimes refer to this as the "integrated porosity" to emphasize its integrated nature. Because the medium is incompressible, we will consider the porosity not as a variable, but rather a parameter, and assume that its fine-scale representation ϕ is known. Thus, the reconstruction operator is not needed, since we assume we know the fine-scale representation exactly, but we can still write it formally as $\hat{\phi} = \mathcal{R}^{II}_\phi \Phi \equiv \phi$.

Because density is constant, we can define coarse-scale "fractional" saturations from the coarse-scale masses based on the following identity:

$$M^i = \Phi \sum_\alpha M^i_\alpha \varrho_\alpha S_\alpha.$$

This implies that for immiscible, incompressible systems, the compression operator for saturation is the weighted vertical average saturation,

$$S_\alpha = \mathcal{C}_{s_\alpha}\boldsymbol{\xi} \equiv \frac{1}{\Phi}\int_{\zeta_B}^{\zeta_T} \phi s_\alpha dx_3.$$

For the case of incompressible fluids and solid and constant mass fractions, we can solve for phase saturations as an explicit function of the coarse mass densities, for example, the CO_2 saturation may be written as

$$S_c = \mathcal{C}_{s_c}\hat{\boldsymbol{\xi}} = \mathcal{C}_{s_c}\mathcal{R}_{M^i}^{II} M^i = \frac{M^i - \Phi M_b^i \varrho_b}{\Phi\left(M_c^i \varrho_c - M_b^i \varrho_b\right)}. \tag{4.12}$$

Saturations are bounded between 0 and 1, and they sum to one when all phases are considered:

$$\sum_\alpha S_\alpha = 1. \tag{4.13}$$

This allows us to eliminate one of the saturations (as was done to derive Eq. 4.12). It is easy to show by substitution into Equation (4.5) that the compression operator for saturation given by Equation (4.12) is equivalent to the porosity-weighted average saturation for the immiscible and incompressible system.

As commented in Section 4.3, pressure decouples in these systems and cannot be reconstructed from the masses in the system. We will therefore adopt (at least) one of the coarse-scale phase pressures as an independent variable, and use the Dupuit-based compression and reconstruction operators $\mathcal{C}_{P_\alpha}$ and $\mathcal{R}_{p_\alpha}^D$.

With this set of definitions, we now have a complete set of compression operators suited to our model simplification, and have also chosen reconstruction operators for density, porosity, mass fraction, and pressure. What remains is to define a reconstruction operator for saturation (the reconstruction operator for mass is equivalent in the constant density case because of the relationship $\widehat{m^i} = \sum_\alpha \widehat{s_\alpha} \rho_\alpha \phi m_\alpha^i$).

To define the reconstruction operator for saturation, we first note that pressure is decoupled from mass density because an equation of state (EoS) relating density and pressure does not appear in incompressible models. However, for the two-phase system we have a relationship between saturation and capillary pressure. Because we are considering a model at the temporal macroscale, both capillary and buoyant forces are assumed to have equilibrated in the vertical direction. This allows us to use the capillary fringe concepts introduced in Chapter 3. At the fine scale, we can write the capillary pressure as a function of saturation as

$$\Delta_\alpha p \equiv p_b - p_c = -p^{cap}(s_c, s_c^t). \tag{4.14}$$

Here we have applied the operator Δ_α to the fine-scale pressure. Note that the definition of Δ_α as the brine value minus the CO_2 value makes the capillary pressure equal to $-\Delta_\alpha p$, assuming brine to be the wetting fluid. We have chosen to use the CO_2 saturation, s_c, and we have also introduced the notation s_c^t to denote the time history of saturation. This allows us to emphasize that the capillary pressure p^{cap} is a function of not only the current value of saturation, s_c, but also the saturation history, thereby allowing for hysteresis in the function. We note an important property of the capillary pressure, which is that for any saturation history s_c^t, the capillary pressure is a monotone function of the current saturation s_c. The function is therefore invertible. As in Section 3.5.2, we refer to the inverted capillary pressure function as the *capillary saturation function*, denoted s^{cap}, which is expressed as follows:

$$s_c = s^{cap}(-\Delta_\alpha p, s_c^t). \tag{4.15}$$

Equation (4.14) leads to a relationship between coarse-scale variables and also provides us with a reconstruction operator. Consider a macroscale pressure difference $\Delta_\alpha P$, and a known time history of the (reconstructed) saturation $\widehat{s_c}^t$. Then, Equation (4.14), together with the definition of $\mathcal{R}_p^D$ given in Equation (4.11), implies the relationship

$$p^{cap}\left(\widehat{s_c}, \widehat{s_c}^t\right) = -\Delta_\alpha p = -\Delta_\alpha \mathcal{R}_p^D P = -\Delta_\alpha P - (\boldsymbol{g}\cdot\boldsymbol{e}_3)\Delta_\alpha \rho(x_3 - \zeta_P).$$

Using Equation (4.15), we can invert this expression to define a reconstruction

$$\widehat{s_c} = s^{cap}\left(-\Delta_\alpha P - (\boldsymbol{g}\cdot\boldsymbol{e}_3)\Delta_\alpha \rho(x_3 - \zeta_P), \widehat{s_c}^t\right) \equiv \mathcal{R}_{s_c}^{II}\{-\Delta_\alpha P\}. \tag{4.16}$$

A simple choice for independent variables in this system is the two coarse-scale phase pressures, from which the fine-scale capillary pressure and saturation can be reconstructed using Equations (4.11) and (4.16). Thereafter, coarse saturations (now both dependent variables) are defined as a function of the coarse pressures through Equation (4.13), and the coarse capillary-saturation function can be defined as,

$$S_c = S^{cap}\left(-\Delta_\alpha P, \widehat{s_c}^t\right) \equiv \mathcal{C}_{S_c}\mathcal{R}_{s_c}^{II}\{-\Delta_\alpha P\}. \tag{4.17}$$

Note that Equation (4.17) is an application of the consistency requirement $\Xi = \mathcal{C}_\Xi \mathcal{R}_\xi \Xi$ discussed in Sections 4.2 and 4.3.

While use of the two phase pressures as independent variables is acceptable, it is often preferable to use one pressure and one phase saturation as the two independent variables, in part because this facilitates mass conservative discretizations at the coarse scale when solving the equations numerically. This choice of variables requires the invertibility of the coarse capillary saturation function, which we analyze as follows.

The function s^{cap} is monotone in $-\Delta_\alpha p$, since it is the inverse of the monotone function p^{cap}. This can be expressed as

$$0 < -\frac{\partial}{\partial \Delta_\alpha p} s^{cap}(-\Delta_\alpha p, s_c^t) = -\frac{\partial}{\partial \Delta_\alpha P} s^{cap}(-\Delta_\alpha P - (\boldsymbol{g}\cdot\boldsymbol{e}_3)\Delta_\alpha \rho(x_3 - \zeta_P), s_c^t).$$

Thus, for any x_3, we have that $\widehat{s_c}$ is a monotone function of $-\Delta_\alpha P$. From this it follows that $S^{cap} = \mathcal{C}_S \widehat{s_c}$ is a monotone function of $-\Delta_\alpha P$, since we know that $\mathcal{C}_s$ is a weighted average. More formally, we have that

$$-\frac{\partial}{\partial \Delta_\alpha P} S^{cap}\left(-\Delta_\alpha P, \widehat{s_c}^t\right)$$
$$= -\frac{\partial}{\partial \Delta_\alpha P}\frac{1}{\Phi}\int_{\zeta_B}^{\zeta_T} \phi s^{cap}\left(-\Delta_\alpha P - (\boldsymbol{g}\cdot\boldsymbol{e}_3)\Delta_\alpha\rho(x_3-\zeta_P), s_c^t\right)dx_3$$
$$= \frac{1}{\Phi}\int_{\zeta_B}^{\zeta_T} \phi\left[-\frac{\partial}{\partial \Delta_\alpha P} s^{cap}\left(-\Delta_\alpha P - (\boldsymbol{g}\cdot\boldsymbol{e}_3)\Delta_\alpha\rho(x_3-\zeta_P), s_c^t\right)\right]dx_3 > 0,$$

where the inequality holds since both terms in the integral are positive.

Therefore, we have that S^{cap} is monotone and invertible in $-\Delta_\alpha P$, and we denote its inverse as the coarse capillary function $P^{cap} = P^{cap}(S_c, \widehat{s_c}^t)$. With this in hand, the reconstruction operator for the capillary pressure can take the coarse saturation as an argument, which we write as follows:

$$\widehat{p^{cap}} = -\widehat{\Delta_\alpha p} = -\mathcal{R}^{II}_{\Delta_\alpha p} S_c = P^{cap}\left(S_c, \widehat{s_c}^t\right) - (\boldsymbol{g}\cdot\boldsymbol{e}_3)\Delta_\alpha\rho(x_3-\zeta_P). \tag{4.18}$$

This allows us to also express the saturation reconstruction defined in Equation (4.16) in terms of the coarse saturation as

$$\widehat{s_c} = \mathcal{R}^{II}_{s_c} S_c = s^{cap}(-\mathcal{R}^{II}_{\Delta_\alpha p} S_c, \widehat{s_c}^t). \tag{4.19}$$

With the reconstructed saturation from Equation (4.19), we now have a complete coarse-scale model for immiscible, constant-density fluids. The model incorporates the mesoscale in a way that is consistent with equilibrium between gravitational and capillary forces. As independent variables, we may choose either the two coarse pressures, or a coarse pressure and a coarse saturation. In either case, our model consists of the conservation of each fluid (Eq. 4.6), coarse fluxes defined through the reconstructed fine scale (Eq. 4.7), and compression and reconstruction operators associated with the immiscible and incompressible assumptions as well as the important Dupuit assumption.

When writing the governing equations in terms of one pressure and one saturation, we obtain an equation analogous to Equation (4.6) in terms of saturation. The resulting mass balance equation for phase α may be written as follows:

$$\Phi\frac{\partial S_\alpha}{\partial t} + \boldsymbol{\nabla}_\parallel\cdot\boldsymbol{U}_\alpha = \Upsilon_\alpha. \tag{4.20}$$

With no flow through the top and bottom boundaries of the formation, Υ_α is defined as

$$\Upsilon_\alpha = \sum_i \frac{M^i_\alpha \Psi^i_\alpha}{\varrho_\alpha}.$$

Using the reconstruction operators, we express the flux as

$$\boldsymbol{U}_\alpha = \int_{\zeta_B}^{\zeta_T} \boldsymbol{e}_\parallel\cdot\boldsymbol{u}_\alpha dx_3 = -\int_{\zeta_B}^{\zeta_T} \boldsymbol{\lambda}_{\alpha,\parallel}\left(\mathcal{R}^{II}_{s\alpha} S_\alpha, \widehat{s_c}^t\right)\boldsymbol{k}_\parallel\left(\boldsymbol{\nabla}_\parallel \mathcal{R}^D_{p\alpha} P_\alpha - \rho_\alpha \boldsymbol{e}_\parallel\cdot\boldsymbol{g}\right)dx_3. \tag{4.21}$$

Here we have again used the assumption that one of the main anisotropy directions of both permeability and mobility is aligned with the x_3-coordinate.

The gradient term in Equation (4.21) takes the form

$$\left(\nabla_{\|}\mathcal{R}_{p\alpha}^{D}P_\alpha - \rho_\alpha e_{\|}\cdot g\right) = \nabla_{\|}P_\alpha + \rho_\alpha(g\cdot e_3)\nabla_{\|}(x_3 - \zeta_P) - \rho_\alpha e_{\|}\cdot g.$$

With the following definitions of coarse permeability, coarse mobilities, and coarse-scale gravity as

$$K = \int_{\zeta_B}^{\zeta_T} k_{\|}dx_3,\quad \Lambda_\alpha\left(S_\alpha, \widehat{s_c}^{\,t}\right) = \int_{\zeta_B}^{\zeta_T} \lambda_{\alpha,\|}\left(\mathcal{R}_{s\alpha}^{II}S_\alpha, \widehat{s_c}^{\,t}\right)k_{\|}dx_3 K^{-1},\quad \text{and}$$

$$G = e_{\|}\cdot g + (g\cdot e_3)\nabla_{\|}\zeta_P,$$

Equations (4.20) and (4.21) combine to yield

$$\Phi\frac{\partial S_\alpha}{\partial t} - \nabla_{\|}\cdot\left(\Lambda_\alpha\left(S_\alpha, \widehat{s_c}^{\,t}\right)K(\nabla_{\|}P_\alpha - \varrho_\alpha G)\right) = \Upsilon_\alpha. \tag{4.22}$$

Equation (4.22) is the coarse-scale saturation equation in terms of only coarse-scale variables under the immiscible, incompressible conditions. We have used the coarse-scale notation for density, although for this incompressible model we will always have that $\varrho_\alpha = \rho_\alpha$. The system is closed by the two constitutive relationships given by $\Sigma_\alpha S_\alpha = 1$ and the macroscale capillary curve $P^{cap}(S_c, \widehat{s_c}^{\,t})$. Notice that the coarse mobility attains a hysteretic nature both from the hysteresis of the fine-scale mobilities and because of the hysteresis in the saturation reconstruction.

Equation (4.22) is a key result in our approach. Most importantly, it achieves our objective of being a coarse-scale equation for CO_2 storage. It has several crucial properties. First, it is derived in a rigorous way, is consistent with coarse-scale constraints (mass balance), and introduces modeling assumptions in a systematic and transparent fashion. Second, all parameters and parameter functions in Equation (4.22) have simple, computable definitions. Third, the derivation allows for heterogeneity of material properties and a general description of the domain. This allows for application to real field cases. Finally, Equation (4.22) has the virtue of having exactly the same form as the fine-scale equation, except for being one dimension lower. Additional complications arise only in the heterogeneous gravity vector. The tensorial nature of mobility is dependent on the heterogeneity of horizontal anisotropy of the geological formation, which in most applications can be expected to be less important than the anisotropy between the vertical and horizontal directions. As such, we expect the coarse equations to exhibit less anisotropy in mobility than the original fine-scale models. Most existing simulation software for porous media flows can be applied directly, or with minor adaptations, for the coarse-scale equations, with the only change being an expanded definition of the gravity term, and the interpretation of the output. We emphasize that while the dependence on the horizontal coordinates has been suppressed throughout the section, the coarse-scale equations will in general inherit similar degrees of spatial heterogeneity to their fine-scale counterparts. Consequently, the coarse-scale parameters are in general heterogeneous.

4.6 BASIC SHARP INTERFACE MODELS

The sharp interface assumption can be obtained as a special case of the immiscible, constant density model of the previous section. Indeed, we will not change the definition of any of the coarse variables from the immiscible and incompressible case, although we will add tertiary variables in order to facilitate the sharp interface description.

Sharp interface models are often appropriate for a temporal mesoscale, before the system has established a capillary fringe, or for systems where the thickness of the capillary fringe is small enough to be neglected. Under either condition, the transition zone in the capillary saturation curve is negligible, so that the capillary saturation may be expressed as follows:

$$s^{cap}\left(-\Delta_\alpha p, \widehat{s_c}^{\,t}\right) = \begin{cases} s_{c,T} & \text{if} \quad -\Delta_\alpha p \geq p_{entry}^{cap} \\ s_{c,B} & \text{if} \quad -\Delta_\alpha p < p_{entry}^{cap} \end{cases}.$$

Here the subscripts B and T continue to denote "bottom" and "top," respectively, while the threshold capillary pressure, p_{entry}^{cap}, represents the entry pressure associated with CO_2 invasion into the brine-filled region. The sharp interface model has a "brine region" below the sharp interface (saturation $s_{c,B}$) and a "mobile CO_2 region" above the interface (saturation $s_{c,T}$), as shown in Figure 4.5. There may also be a third region, depending on the flow history, which we refer to as the "residual CO_2 region." For now, we assume that the residual CO_2 region can be omitted. We will discuss the residual CO_2 region at the end of this section. Note that for the usual case of CO_2 injection into an initially brine-filled formation, the CO_2 saturation below the sharp interface will be zero, and above the interface it will be one minus the residual brine saturation: $s_{c,B} = 0$ and $s_{c,T} = 1 - s_b^{res}$.

As illustrated in Figure 4.5, we denote the surface representing the lower extent of the mobile region by the function ζ_M, with subscript M for mobile, such that the interface is given by $x_3 = \zeta_M(x_1, x_2, t)$. This interface location is a macroscale tertiary

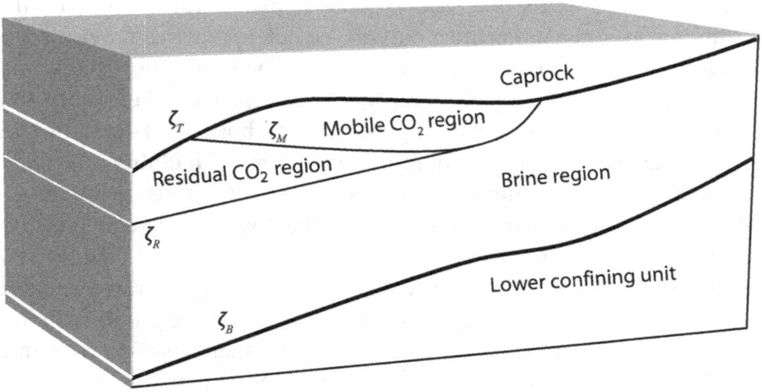

Figure 4.5 Schematic of a sharp interface description of the CO_2–brine system.

variable, specific to this modeling choice. It also corresponds directly to the surface along which the pressure difference between the phases is equal to the capillary entry pressure. We can reconstruct the fine-scale saturation from ζ_M through its definition

$$\widehat{s_\alpha} = \mathcal{R}^{SI}_{s_\alpha}\zeta_M = \begin{cases} s_{\alpha,B} & \text{if} \quad x_3 \le \zeta_M \\ s_{\alpha,T} & \text{if} \quad x_3 > \zeta_M \end{cases}.$$

Note that we use superscript *SI* to denote the sharp interface reconstruction. Further, the coarse saturation is a function of ζ_M through application of the compression operator applied to the reconstructed fine-scale saturation profile, $\widehat{s_\alpha}$, such that $S_\alpha = \mathcal{C}_{S_\alpha}\mathcal{R}^{SI}_{s_\alpha}\zeta_M$. When porosity is independent of x_3, this relationship simplifies and can be made explicit through the following relationship:

$$S_\alpha = \frac{(\zeta_M - \zeta_B)s_{\alpha,B} + (\zeta_T - \zeta_M)s_{\alpha,T}}{\zeta_T - \zeta_B} = \frac{h_{M,B}s_{\alpha,B} + h_{T,M}s_{\alpha,T}}{h_{T,B}} = s_{\alpha,B} + \frac{h_{T,M}}{h_{T,B}}(s_{\alpha,T} - s_{\alpha,B}),$$

where we recall the notation h to represent the thickness of different regions, using the convention that the subscripts denote the top and bottom surfaces for the given region. This means that $h_{M,B} = \zeta_M - \zeta_B$ and $h_{T,M} = \zeta_T - \zeta_M$, and also implies that $h_{T,M} + h_{M,B} = h_{T,B}$. It is also useful to express the thickness of the mobile region in terms of the coarse saturations,

$$h_{T,M} = \frac{S_\alpha - s_{\alpha,B}}{s_{\alpha,T} - s_{\alpha,B}} h_{T,B}.$$

For sharp interface models, thickness of the CO_2-rich layer $h_{T,M}$, is frequently chosen as an independent variable. We can substitute these relationships into (both of) the coarse saturation equations (4.22) to get two equations in terms of the CO_2 thickness:

$$\frac{\Phi(s_{\alpha,T} - s_{\alpha,B})}{h_{T,B}}\frac{\partial h_{T,M}}{\partial t} - \nabla_\| \cdot (\mathbf{\Lambda}_\alpha(h_{T,M})\boldsymbol{K}(\nabla_\| P_\alpha - \varrho_\alpha \boldsymbol{G})) = \Upsilon_\alpha. \tag{4.23}$$

In this expression, we can evaluate the coarse mobility as

$$\mathbf{\Lambda}_\alpha(h_{T,M}) = \int_{\zeta_B}^{\zeta_T - h_{T,M}} \boldsymbol{\lambda}_{\alpha,\|}(s_{\alpha,B})\boldsymbol{k}_\| dx_3 \boldsymbol{K}^{-1} + \int_{\zeta_T - h_{T,M}}^{\zeta_T} \boldsymbol{\lambda}_{\alpha,\|}(s_{\alpha,T})\boldsymbol{k}_\| dx_3 \boldsymbol{K}^{-1}.$$

Note that we have no hysteresis in this simplest sharp interface model. When the permeability and mobility are independent of the x_3-coordinate, the expression for coarse mobility simplifies further, and becomes linear in the CO_2 thickness:

$$\mathbf{\Lambda}_\alpha(h_{T,M}) = \boldsymbol{\lambda}_\alpha(s_{\alpha,B})\frac{(h_{T,B} - h_{T,M})}{h_{T,B}} + \boldsymbol{\lambda}_\alpha(s_{\alpha,T})\frac{h_{T,M}}{h_{T,B}}. \tag{4.24}$$

Similarly, the macroscale capillary pressure function can be obtained directly, since by definition of the fine-scale capillary saturation function and the sharp interface, $\Delta_\alpha \hat{p}(x_3 = \zeta_M) = -p^{cap}_{entry}$; thus, the pressure difference at the datum is

$$\Delta_\alpha P = \Delta_\alpha \mathcal{C}_P \hat{p} = \Delta_\alpha \hat{p}(x_3 = \zeta_P) = -p^{cap}_{entry} - (\boldsymbol{g} \cdot \boldsymbol{e}_3)\Delta_\alpha \rho(h_{T,P} - h_{T,M}). \tag{4.25}$$

We see that the macroscale capillary pressure function is linear with respect to ζ_M, and represents the effects of gravity on the interface between CO_2 and brine. It is most often the case that only one fluid is mobile on each side of the interface (brine below and CO_2 above). In particular, we have that $s_{c,B} = \lambda_c(s_{c,B}) = \lambda_b(s_{b,T}) = 0$.

Equation (4.23), together with the coarse mobility (Eq. 4.24) and coarse capillary pressure (Eq. 4.25), constitute the usual sharp interface equations for a porous medium with finite vertical extent and negligible variation in parameters along the x_3-coordinate. We see that these coarse-scale equations have the same structure as the fine-scale equations, but with linear mobility and capillary pressure, even if the fine-scale equivalents are nonlinear.

A particular feature of this sharp interface model is the analytical evaluation of the integral in the global pressure formulation due to the linear constitutive functions. From Equation (3.25) in Section 3.6 we obtain for isotropic mobility functions

$$\begin{aligned}\frac{P_G - P_b}{(\boldsymbol{g}\cdot \boldsymbol{e}_3)\Delta_\alpha \rho h_{T,B}} &= \frac{1}{(\boldsymbol{g}\cdot \boldsymbol{e}_3)\Delta_\alpha \rho h_{T,B}} \int_0^{S_c} \frac{\Lambda_c(S_c')}{\sum_\alpha \Lambda_\alpha(S_\alpha')} \frac{dP^{cap}(S_c')}{dS_c'} dS_c' \\ &= \frac{\lambda_c}{\Delta_\alpha \lambda}\left[\frac{S_c}{s_c} + \frac{\lambda_b}{\Delta_\alpha \lambda} \ln\left(1 - \frac{S_c}{s_c}\frac{\Delta_\alpha \lambda}{\lambda_b}\right)\right]\end{aligned} \tag{4.26}$$

where we have used the shorthand $\lambda_c = \lambda_c(s_{c,T})$, $\lambda_b = \lambda_b(s_{b,B})$ and the convention of the Δ_α operator with $\Delta_\alpha \lambda = \lambda_b - \lambda_c$. Note that $\Delta_\alpha \lambda$ will usually be negative. We will later see that since the global pressure is the driving force of the total flow, it naturally appears in the analysis of CO_2 injection. Having an explicit expression relating the global pressure to the coarse-scale fluid pressures is therefore convenient. Note in particular that when $S_c = 0$, our choice of lower integration limit in the definition of the global pressure ensures that it is equal to the coarse-scale brine pressure.

We close this section by introducing a simple hysteresis model in the context of a sharp interface approximation (general hysteresis for immiscible, incompressible systems is covered in the previous section). We distinguish between drainage and imbibition, so that we obtain three regions: brine ahead of the CO_2 plume; brine imbibing into the CO_2 region, with CO_2 at residual saturation; and CO_2 draining the brine, with brine at residual saturation. These regions are separated by two interfaces ζ_M and ζ_R, with the former still denoting the lowermost extent of the mobile CO_2 region, and the latter bounding the residual CO_2 region, as illustrated in Figure 4.5. A saturation reconstruction (indicated by superscript *HSI* for hysteretic sharp interface) honoring this hysteresis model can be expressed as

$$\widehat{s_\alpha} = \mathcal{R}_{s_\alpha}^{HSI}\zeta = \begin{cases} s_{\alpha,B} & \text{if} \quad x_3 \le \zeta_R \\ s_{\alpha,R} & \text{if} \quad \zeta_R < x_3 \le \zeta_M . \\ s_{\alpha,T} & \text{if} \quad x_3 > \zeta_M \end{cases}$$

Since we have introduced a new tertiary variable ζ_R, we need to supplement our equation with an additional constitutive assumption. It is natural to define the boundary of the residual CO_2 region as the furthest extent (over time) of the mobile CO_2 region, thus

$$\zeta_R = \min_{t' \in [o,t]} \zeta_M(t'). \tag{4.27}$$

Consistent with the general hysteresis model from Section 4.5, ζ_R can be interpreted as a parameterization of the history of the problem.

We can now write down the equations for sharp interface models with hysteresis, using the simplification that porosity and permeability are independent of x_3. The appropriate equations, derived from Equation (4.22) in the same manner as Equation (4.23), are given as

$$\frac{\Phi\Delta^* s_\alpha}{h_{T,B}} \frac{\partial h_{T,M}}{\partial t} - \nabla_{\|} \cdot \left(\mathbf{\Lambda}_\alpha (h_{T,M}, h_{M,R}) \boldsymbol{K} (\boldsymbol{\nabla}_{\|} P_\alpha - \varrho_\alpha \boldsymbol{G}) \right) = \Upsilon_\alpha, \tag{4.28}$$

where $\Delta^* s_\alpha$ denotes the jump in saturation across the interface, which is

$$\Delta^* s_\alpha = \begin{cases} s_{\alpha,T} - s_{\alpha,B} & \text{if} \quad h_{M,R} = 0 \\ s_{\alpha,T} - s_{\alpha,R} & \text{if} \quad h_{M,R} > 0 \end{cases}.$$

This implies that we consider two cases: when $h_{M,R} = 0$ we have primary drainage, as seen at the leading front of the CO_2 plume in Figure 4.5; conversely, when $h_{M,R} > 0$ we have either imbibition or secondary drainage at ζ_M. We also note that when permeability and mobility functions are independent of x_3, the coarse mobility is still linear in both its arguments:

$$\mathbf{\Lambda}_\alpha (h_{T,M}, h_{M,R}) = (h_{T,B} - h_{T,M} - h_{M,R}) \boldsymbol{\lambda}_{\alpha,\|}(s_{\alpha,B}) + h_{M,R} \boldsymbol{\lambda}_{\alpha,\|}(s_{\alpha,R}) + h_{T,M} \boldsymbol{\lambda}_{\alpha,\|}(s_{\alpha,T}).$$

The coarse capillary pressure function remains unchanged from the non-hysteretic formulation, when expressed in terms of the thickness $h_{T,M}$. This leads to an interesting observation: since coarse saturation is a function of both ζ_M and ζ_R, capillary pressure cannot be expressed uniquely in terms of only coarse saturation—it becomes hysteretic with respect to this variable. Thus, in the hysteretic sharp interface model, we see that the complexity of the capillary pressure function (i.e., if it is hysteretic in nature) is dependent on the choice of coarse variables. This is an example of the more general observation that the complexity of coarse models we derive is necessarily dependent on our choice of coarse representation. The optimal coarse representation is in general not obvious a priori, a fact that is important to keep in mind when embarking on multiscale modeling of complex systems (see Box 4.1).

The present model and discussion concludes the section on "Basic Sharp Interface Models." In the remainder of the chapter, we will present three rather independent generalizations of the immiscible, incompressible models presented so far. These allow us to consider non-Dupuit pressure variation, compressible systems, and dissolution with convective mixing. These generalizations all build on what we have covered so far, but do not build on each other, and can therefore be read in any order.

BOX 4.1 *Constitutive Functions and Coarse Variables*

The simple forms of the coarse constitutive functions that appear in sharp interface formulations allow us to better understand the implications of our choice of coarse pressure on the constitutive functions in our models. Our coarse pressure is defined as the fine-scale pressure at a datum, or sampling height, $\zeta_P = \zeta_P(\boldsymbol{x}_\parallel, t)$. There are three natural choices suggested by the system geometry: a flat surface; an internal surface, such as ζ_M; or the top (or bottom) formation boundary. We will briefly review these options here.

A flat surface (either horizontal or tilted) sampling height is defined as $\zeta_P = \boldsymbol{a} \cdot \boldsymbol{x}_\parallel + b$, for some vector $\boldsymbol{a}$ and scalar $\boldsymbol{b}$. When considering the definition of $\boldsymbol{G}$ given in Section 4.5, we see that a flat surface leads to a constant gravitational vector. This has advantages when using vertically integrated formulations in traditional codes, since no modification needs to be considered to account for spatially dependent gravity. However, we see from Equation (4.25) that a flat sampling height leads to a spatially variable coarse capillary pressure, even for homogeneous formations of constant thickness.

Choosing the internal surface is suggested by trying to simplify the coarse capillary pressure. Since $\zeta_P = \zeta_M$ implies that $h_{T,P} = h_{T,M}$, we see (again from Eq. 4.25) that this choice eliminates everything but the fine-scale entry pressure from the coarse capillary pressure. The fine-scale entry pressure will frequently be negligible or constant, so that coarse capillary pressure becomes particularly simple. Unfortunately, this choice implies that the gravitational vector becomes dependent on ζ_M, and in other words, temporally varying and dependent on gradients of the solution. Thus, the coarse gravitational vector now has a much more complex character.

Finally, one may choose the sampling height to align with, for example, the top formation boundary, such that $\zeta_P = \zeta_T$. We see that the coarse gravitational vector is now essentially the gradient of the top surface, which is consistent with the interpretation of $\boldsymbol{G}$ as a coarse driving force. Further, the capillary pressure becomes geometry-independent, since the term $h_{T,P} = 0$ in Equation (4.25), and is only a function of the solution. Due to this splitting of geometry and solution variables, which is consistent with the role of these terms on the fine scale, we consider setting the sampling height equal to a formation boundary as the natural choice for defining our coarse variable.

4.7 VERTICALLY STRUCTURED PRESSURE RECONSTRUCTION

The previous sections have highlighted how we can use different saturation reconstructions to create models of different complexity, both in terms of physics and in terms of mathematical nonlinearities. In this section we focus on the pressure reconstruction, which to this point has been based on the Dupuit approximation. We now consider a reconstruction that may be used when the Dupuit assumption is deemed unsatisfactory. Such is the case when the interfaces of the problem (top and bottom of the aquifer, as well as the fluid interface in a sharp interface model) form surfaces whose normal vector is not close to $\boldsymbol{e}_3$, or when the flow rates in the $\boldsymbol{e}_3$ direction are non-negligible. Significant flows in the $\boldsymbol{e}_3$ direction often occur near wells, particularly pumping or leaky wells. In these cases, we can observe a significant component $\boldsymbol{u}_\alpha \cdot \boldsymbol{e}_3$, such that the Dupuit assumption, Equation (4.10), is no longer an adequate approximation. We note that one of the significant advantages of the multiscale framework is the flexibility to consider the underlying assumptions of our models independently of one another, with the focus herein being on pressure reconstruction.

A natural generalization of the Dupuit assumption is to consider a simple low-order approximation to $\boldsymbol{u}_\alpha$ along the $\boldsymbol{e}_3$ direction. Let us first consider what happens near a stationary interface ζ such as the top or bottom boundary. Let the unit normal vector be defined as $\boldsymbol{v}_\zeta$. If the interface is impermeable, then at the interface we have a condition of zero normal flux:

$$0 = \boldsymbol{u}_\alpha^\pm \cdot \boldsymbol{v}_\zeta = \boldsymbol{u}_\alpha^\pm \cdot \boldsymbol{e}_3 - \left(\boldsymbol{u}_\alpha^\pm \cdot \boldsymbol{e}_\parallel\right) \cdot \nabla_\parallel \zeta = u_{\alpha,3}^\pm - \boldsymbol{u}_{\alpha,\parallel}^\pm \cdot \nabla_\parallel \zeta.$$

Thus, we can express the x_3-component of the flow based on the components in the remaining directions and the gradient of the interface. If $\nabla_\parallel \zeta = 0$, then we simply obtain that $u_{\alpha,3}^\pm = 0$. We note that since the flow parallel to an interface is not continuous across the interface, the x_3-component in general takes different values on each side of the interface, denoted by $u_{\alpha,3}^+$ for above the interface, and $u_{\alpha,3}^-$ for below the interface. The corresponding expression near a time-dependent interface, with a flow rate $v_\alpha = \boldsymbol{u}_\alpha \cdot \boldsymbol{v}_\zeta$ across the interface, requires addition of both the flow across the interface and the rate of movement of the interface, which we recall (from the interface conditions discussed in Section 2.3.3) leads to expressions of the form

$$u_{\alpha,3}^\pm = \pm v_\alpha + \boldsymbol{u}_{\alpha,\parallel}^\pm \cdot \boldsymbol{\nabla}_\parallel \zeta + \Delta^* s_\alpha \phi \frac{\partial \zeta}{\partial t}, \tag{4.29}$$

where $\Delta^* s_\alpha$ represents the change in saturation across the interface. We can now construct a linear interpolation of $u_{\alpha,3}$ between two interfaces ζ_1 and ζ_2, with $\zeta_2 > \zeta_1$:

$$u_{\alpha,3} \approx u_{\alpha,3}^+\Big|_{\zeta_1} + \left(u_{\alpha,3}^-\Big|_{\zeta_2} - u_{\alpha,3}^+\Big|_{\zeta_1}\right) \cdot \frac{x_3 - \zeta_1}{\zeta_2 - \zeta_1}.$$

If we use this expression for $\boldsymbol{u}_\alpha \cdot \boldsymbol{e}_3$ in Equation (4.10) (rather than setting it to zero), we can derive the first-order vertically structured pressure reconstruction, denoted by superscript *VS*, through rearrangement and subsequent integration of the pressure term:

$$\widehat{p_\alpha} = \mathcal{R}_{p\alpha}^{VS} \Xi = P_\alpha + (\boldsymbol{g} \cdot \boldsymbol{e}_3) \int_{\zeta_p}^{x_3} \rho_\alpha dx_3' - \int_{\zeta_p}^{x_3} \frac{u_{\alpha,3}}{\lambda_{\alpha,3}\left(\widehat{s_\alpha}\right) k_3} dx_3'. \tag{4.30}$$

Here we have used the general coarse-scale solution vector Ξ. Contrary to the Dupuit reconstruction, the vertically structured reconstruction of pressure depends not only on the coarse phase pressure, but in general depends on more coarse-scale information because a saturation reconstruction is needed both implicitly in the approximation of $u_{\alpha,3}$ and for the evaluation of $\lambda_{\alpha,3}\left(\widehat{s_\alpha}\right)$. There is also an implicit dependence on the reconstructed pressure itself, since the evaluation of Equation (4.29) requires fluxes that are defined using Darcy's law together with the reconstructed fluxes. And

Equation (4.30) introduces the permeability along the x_3-direction, which is absent from other reconstructions.

The significant added complications of using vertically structured reconstructions as opposed to the simple Dupuit reconstruction of pressure explains why the latter is almost always applied. As reviewed in the beginning of this section, the vertically structured reconstruction is of primary interest near pumping and leaking wells; analytical approximations in this case will be presented in Section 5.1.2.

4.8 COARSE EQUATION OF STATE

In order to understand how to incorporate density changes into coarse models, we now generalize the immiscible models from Section 4.5. Our concern in this section will be the compression operator for density, and the construction of consistent reconstruction operators for both pressure and density jointly. The reconstruction operators defined in this section will be denoted by a superscript ϱ to denote models that include variable density. The compression operators will use a similar subscript notation.

We recall from Equation (3.8) that for immiscible systems, mass of each phase is conserved, and is given by $\rho_\alpha \phi s_\alpha$. We want to retain this notion on the coarse scale, and therefore require that the product $\varrho_\alpha \Phi S_\alpha$ represents the mass of the phase, and is consistent with the compression of mass given in Section 4.3. For this to hold, the compression of density must be defined as

$$\varrho_\alpha \equiv \mathcal{C}_{\varrho_\alpha}\boldsymbol{\xi} \equiv \frac{\int_{\zeta_B}^{\zeta_T} \rho_\alpha \phi s_\alpha dx_3}{\mathcal{C}_\Phi \boldsymbol{\xi} \cdot \mathcal{C}_{S_\alpha}\boldsymbol{\xi}} = \frac{\int_{\zeta_B}^{\zeta_T} \rho_\alpha \phi s_\alpha dx_3}{\Phi S_\alpha}.$$

Note that since our coarse-scale porosity has units of length, coarse-scale density has the same units as its fine-scale counterpart. However, since integration does not commute with nonlinear functions, we see that

$$\varrho_\alpha = \mathcal{C}_{\varrho_\alpha}\boldsymbol{\xi} \neq \rho_\alpha(\mathcal{C}_{P_\alpha}\boldsymbol{\xi}) = \rho_\alpha(P_\alpha).$$

In other words, the coarse-scale equation of state (EoS) is not equal to the fine-scale EoS. This aspect is frequently neglected in macroscale modeling of porous media.

As was pointed out in Section 4.2, we can obtain a coarse-scale relationship by the successive application of reconstruction and compression. Formally, if we take density to be a dependent variable, the coarse EoS is expressed as

$$\varrho_\alpha = \mathcal{C}_{\varrho_\alpha}\mathcal{R}_{\xi}\Xi,$$

where we recall that Ξ defines the coarse solution vector. Note that we did not specify the reconstruction, and indeed, the coarse-scale EoS is dependent on the reconstruction. This also implies that our definition of a coarse EoS is an implicit definition, since all the pressure reconstructions we have considered are themselves functions of the (reconstructed) density.

Let us consider these ideas in the context of the Dupuit reconstruction, and assume that we have chosen the two phase pressures P_α as our independent coarse variables. Then, enforcing the fine-scale EoS, written as $\rho_\alpha(p_\alpha)$ together with the

Dupuit reconstruction gives the following equations for the reconstructed pressure of each phase:

$$\widehat{p_\alpha} = \mathcal{R}_{p_\alpha}^{D,\rho} P_\alpha \equiv P_\alpha + (\boldsymbol{g} \cdot \boldsymbol{e}_3) \int_{\zeta_P}^{x_3} \rho_\alpha \left(\widehat{p_\alpha} \right) dx_3'.$$

This equation can be integrated directly, after which the reconstruction of density is given simply as

$$\widehat{\rho_\alpha} = \mathcal{R}_{p_\alpha}^{D,\rho} P_\alpha \equiv \rho_\alpha \left(\mathcal{R}_{p_\alpha}^{D,\rho} P_\alpha \right).$$

Similarly, we can reconstruct saturation using the capillary saturation function s^{cap} as we did in Section 4.5. With both a reconstruction of saturation and density, we have sufficient reconstructed information to evaluate our constitutive functions $\varrho_\alpha(P_c, P_b) = \mathcal{C}_{\varrho_\alpha} \mathcal{R}_\xi^{D,\rho} \{P_c, P_b\}$ and $S_\alpha(P_c, P_b) = \mathcal{C}_{S_\alpha} \mathcal{R}_\xi^{D,\rho} \{P_c, P_b\}$.

We make a few comments about these constitutive relationships. First, we note that not only is the coarse-scale EoS in general different from the fine-scale EoS, it also has an expanded functional dependency. This comes from the fact that as pressure and density vary through the vertical direction, the saturation also varies. Thus, thinking of fine-scale saturation as a weighing function in the compression operator for density, we see that saturation leads to preferential weighing of parts of the domain. This is in contrast to the microscale, or ℓ_{lab}, we discussed in Chapters 2 and 3, where we generally assume that the fluids are evenly distributed on the subscale. Second, we note that the interrelationship between saturation and density is the reason why the coarse saturation S_α is no longer a function only of the pressure difference $-\Delta_\alpha P$, as was the case for incompressible models in Section 4.5, but now becomes a function of both pressures independently. By a change of variables, we may still write, for example, $S_\alpha = S_\alpha(-\Delta_\alpha P, P_b)$, which gives us the interpretation that on the coarse scale, the capillary saturation function is dependent not only the pressure difference, but also on the absolute magnitude of pressure.

Using the reconstructions of this section, a complete coarse model for compressible flow, analogous to Equation (3.8), is given by the mass conservation equations written as

$$\frac{\partial}{\partial t}(\varrho_\alpha \Phi S_\alpha) + \nabla_{\parallel} \cdot \boldsymbol{F}_\alpha = \Psi_{\Sigma,\alpha}$$

together with the fluxes

$$\boldsymbol{F}_\alpha = \boldsymbol{e}_{\parallel} \cdot \int_{\zeta_B}^{\zeta_T} \widehat{\rho_\alpha} \boldsymbol{u}_\alpha dx_3.$$

As usual, the fine-scale Darcy flux is evaluated with the reconstructed variables.

The coarse-scale model for flow derived in this section can be linearized in a completely analogous way as in Chapters 2 and 3 to obtain an equation involving the fluid compressibilities C_α.

In closing, we note that for systems where there is relatively little vertical variation in density, it will be a natural approximation to simply use the fine-scale EoS on the coarse scale. However, for aquifers where the injection depth is such that CO_2 is close to the critical point, significant variations in the vertical density profile may occur, and the fine-scale EoS may no longer be adequate.

4.9 SHARP INTERFACE WITH COMPONENT TRANSFER

Our final coarse-scale model addresses the issue of dissolution and convective mixing. We recall from our analysis of length and time scales that these processes, particularly convective mixing, can become important during the macro temporal scale, and will definitely be important at the mega temporal scale.

Although convective mixing happens on longer time scales than the establishment of a capillary front, we will nevertheless use a sharp interface approach to characterize the system. While this may not be fully justified, the combined convective mixing and capillary fringe system is not yet fully understood on the microscale, and we thus lack the basis to undertake macroscale modeling. This approach is also reasonable when the capillary fringe is negligibly small. We model the system by considering the fluids as incompressible, and all phase compositions as at equilibrium at the microscales. For pressure, we will use the Dupuit reconstruction, and neglect the effect of capillary forces on the pressure distribution, as in the more basic sharp interface models.

We denote by superscript *CM* (convective mixing) the reconstruction operators defined in this section. We make further notational simplifications in that we do not consider a region of dry CO_2, which would exist in the absence of residual saturation of brine. This region can be introduced in a straight-forward manner; however, its impact on macroscale measures is limited, so we omit it to simplify the presentation. We also simplify notation by neglecting vertical variation in porosity.

To represent the system on the macroscale, we use the following tertiary variables illustrated in Figure 4.6: the interface between original brine and brine with dissolved CO_2 is ζ_C; the interface between brine and a region where CO_2 is at residual saturation is ζ_R; the interface between the region where CO_2 is mobile (brine at residual saturation) and the region where brine is mobile is ζ_M. The interfaces will necessarily satisfy this ordering, so that $\zeta_B \leq \zeta_c \leq \zeta_R \leq \zeta_M \leq \zeta_T$. Within each region, we take the fluid properties to be constant, and all fine-scale processes are at equilibrium. This concept forms the basis of our reconstructions, and is summarized in the following equations:

$$\widehat{s_c} = \mathcal{R}_s^{CM}\Xi = \begin{cases} 0 & \text{if} \quad \zeta_B < x_3 \leq \zeta_R \\ s_{c,R} & \text{if} \quad \zeta_R < x_3 \leq \zeta_M \\ s_{c,T} & \text{if} \quad \zeta_M < x_3 \leq \zeta_T \end{cases}.$$

These saturations, by definition, have the properties that $\lambda_c(s_{c,R}) = 0$, and $\lambda_b(1 - s_{c,T}) = 0$.

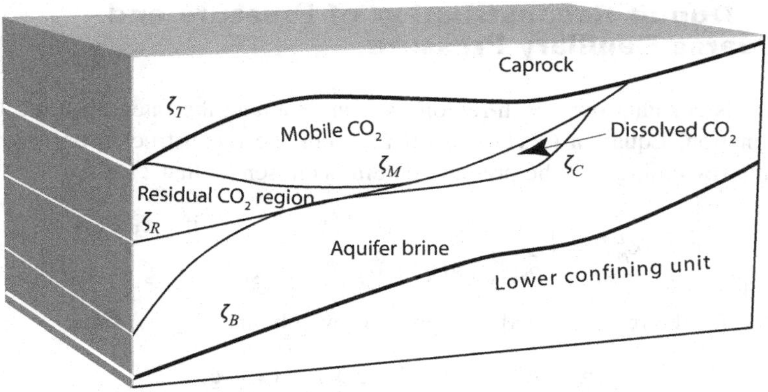

Figure 4.6 Schematic of a sharp interface description of the CO_2–brine system including upscaled convective mixing.

Based on the reconstruction of saturation, and the fact that we only consider equilibrium partitioning at the microscale, we obtain from our coarse-scale variables a reconstruction of mass fractions

$$\widehat{m_\alpha^i} = \mathcal{R}_{m_\alpha^i}^{CM}\Xi = \begin{cases} m_\alpha^{i,eq} & \text{if} \quad \zeta_C \le x_3 \le \zeta_T \\ 1 & \text{if} \quad \alpha = b \text{ and } i = H_2O \text{ and } \zeta_B \le x_3 < \zeta_C \\ 0 & \text{if} \quad \alpha = b \text{ and } i = CO_2 \text{ and } \zeta_B \le x_3 < \zeta_C \end{cases}.$$

From this we recognize that the phase densities can only attain two values, one for mixtures, ρ_α^{mix} and one for pure fluids ρ_α^{pure}. We formalize this as

$$\widehat{\rho_\alpha} = \mathcal{R}_{\rho_\alpha}^{CM}\Xi = \begin{cases} \rho_\alpha^{mix} & \text{if} \quad \zeta_C \le x_3 \le \zeta_T \\ \rho_b^{pure} & \text{if} \quad \alpha = b \text{ and } \zeta_B \le x_3 < \zeta_C \\ \rho_c^{mix} & \text{if} \quad \alpha = c \text{ and } \zeta_B \le x_3 < \zeta_C \end{cases}.$$

Recall that we denote the fine-scale solution vector as $\boldsymbol{\xi}$, and we denote the compound reconstruction operator simply as $\hat{\boldsymbol{\xi}} = \mathcal{R}_\xi^{CM}\Xi$. In addition to the three variable interfaces, we will also employ two coarse-scale phase pressures as independent variables in this section.

Our tertiary variables satisfy the two constraints imposed by consistency in reconstruction and compression; $M^i = \mathcal{C}_{M^i}\mathcal{R}_\xi^{CM}\boldsymbol{\zeta}$, where M^i are the coarse-scale component masses defined in Section 4.3, which propagate according to the mass conservation equation, and the vector $\boldsymbol{\zeta}$ represents the vector of interfaces in the problem. Because we have five independent variables, we must obtain three additional equations based on the microscale. Two of these will be familiar, as they take the role of the coarse capillary pressure and a hysteresis model. The remaining model will be novel to this section, and represents the dynamics of dissolution.

4.9.1 Dupuit Reconstruction of Pressure and the Coarse Capillary Pressure

Since $\widehat{\rho_\alpha}$ is constant over each region, we can evaluate the integral in the Dupuit approximation, Equation (4.11), analytically. For the reconstructed brine pressure we obtain (assuming that the pressure datum is chosen so that $\zeta_P \leq \zeta_C$)

$$\widehat{p_b} = \mathcal{R}_{p_b}^D P_b = \begin{cases} P_b + (\boldsymbol{g}\cdot\boldsymbol{e}_3)\cdot\rho_b^{pure}(x_3-\zeta_P) & \text{if} \quad x_3 \leq \zeta_C \\ \widehat{p_b}(\zeta_C) + (\boldsymbol{g}\cdot\boldsymbol{e}_3)\cdot\rho_b^{mix}(x_3-\zeta_C) & \text{if} \quad \zeta_C < x_3 \end{cases}.$$

Similarly, for the reconstructed CO_2 pressure we obtain

$$\widehat{p_c} = \mathcal{R}_{p_c}^D P_c = P_c + (\boldsymbol{g}\cdot\boldsymbol{e}_3)\cdot\rho_c^{mix}(x_3-\zeta_P).$$

Note that we have extended the definition of reconstructed CO_2 pressure below the region of CO_2 using the density of the mixed CO_2 phase. This expression is constrained to match the appropriate pressure (either brine pressure or the capillary entry pressure) at $x_3 = \zeta_M$. If the entry pressure is used, then we require that $\widehat{p_c}(\zeta_M) = \widehat{p_b}(\zeta_M) + p_{entry}^{cap}$, and from this we obtain the coarse capillary pressure function

$$P^{cap}(\zeta) = p_{entry}^{cap} + (\boldsymbol{g}\cdot\boldsymbol{e}_3)\cdot\left[\Delta_\alpha\rho^{mix}(\zeta_M-\zeta_C) + \left(\rho_b^{pure} - \rho_c^{mix}\right)(\zeta_C-\zeta_P)\right]. \quad (4.31)$$

While we have emphasized the Dupuit reconstruction, due to its simplicity, it is also possible to use the vertically structured reconstruction of pressure if vertical flows become important.

4.9.2 Hysteresis Model

In Section 4.6, we had a simple hysteresis model for ζ_R, given by Equation (4.27). However, since we now include dissolution, we have to account for both convective mixing originating at the bottom of the residual CO_2 region, as well as the possibility of fresh brine flowing into the residual region and dissolving CO_2 as the phases equilibrate. Because of these complexities in the residual zone, it is more convenient to consider writing a mass balance equation associated with the mobile interface ζ_M, when it is distinctly above the residual region. Then, it will not be affected by dissolution, and it satisfies Equation (4.28), which we rewrite for the CO_2 phase in a form that focuses on the temporal dynamics of the interface location:

$$\frac{\partial \zeta_M}{\partial t} = h_{T,B}\frac{\nabla_\| \cdot \boldsymbol{U}_c - \Upsilon_c}{\Phi\left(s_{c,T} - s_{c,R}\right)} \equiv \mathcal{D}_M \Xi.$$

Here we have defined the operator $\mathcal{D}_M$ as the "evolution operator" for ζ_M, which is an expression involving coarse-scale fluxes and source/sink terms. The evolution given by $\mathcal{D}_M$ is valid as long as $\zeta_R < \zeta_M < \zeta_T$, outside of which the interface is constrained by the ordering of the interfaces. This allows us to specify our hysteresis model with the following equation:

$$\frac{\partial \zeta_M}{\partial t} = \begin{cases} \min(\mathcal{D}_M, 0) & \text{if} \quad \zeta_M = \zeta_T \\ \mathcal{D}_M & \text{if} \quad \zeta_R < \zeta_M < \zeta_T . \\ \max\left(\mathcal{D}_M, \dfrac{\partial \zeta_R}{\partial t}\right) & \text{if} \quad \zeta_M = \zeta_R \end{cases} \tag{4.32}$$

We see that we have a system of time-dependent differential equations, constrained by inequalities. This is of importance for implementation, as when interfaces coincide (as they will in parts of the domain), the differential part of the equations are no longer independent. We see that such systems arise naturally in the setting of (multiple) sharp interface models, in order to satisfy both a transport equation as well as the constraint on the ordering of the interfaces. This generalizes the similar situation that occurs in fine-scale models for flow in porous media, which also are constrained by the condition that all saturations must be positive.

4.9.3 CO_2–Saturated Brine

We find it simpler to write an expression for the rate of dissolution of CO_2 into brine than to write an expression for any of the remaining regions. The total amount of dissolved CO_2 in the brine phase is given as the following dependent tertiary variable:

$$\mathcal{M}_b^c \equiv \phi \rho_b^{mix} m_b^{c,eq} \cdot \left[h_{R,C} + s_{b,R} h_{M,R} + s_{b,T} h_{T,M} \right].$$

The mass transfer rate is the subject of our modeling and consists of two parts. In the absence of convective mixing, dissolved brine is produced when pure brine moves into a region of CO_2 (or opposite). This will be the case when there is no convective mixing zone, $\zeta_C = \zeta_R$. When this equality holds, the system of equation has essentially one less unknown, and the resulting model correctly accounts for the dissolution of CO_2 into residual brine through the mass balance equations. Once the interfaces are separate, convective mixing will occur at a rate we presume known and which we denote using the double subscript on Ψ, consistent with the mass transfer notation introduced in Section 3.5. We emphasize the fundamental assumption that the upscaled mass transfer can be expressed in terms of our coarse variables, and write $\Psi_{\mathcal{M}_b^c,\beta}\{\Xi\}$. When $\zeta_B < \zeta_C < \zeta_R$, we then state the balance equation for dissolved CO_2 as

$$\frac{\partial \mathcal{M}_b^c}{\partial t} + \nabla_{\parallel} \cdot \left(\rho_b^{mix} m_b^{c,eq} \int_{\zeta_C}^{\zeta_M} \widehat{\boldsymbol{u}_b} dx_3 \right) = \Psi_{\mathcal{M}_b^c,\beta}\{\Xi\}.$$

In order to honor the ordering constraint on ζ_C, we rewrite this equation explicitly in terms of this variable using the definition of $\mathcal{M}_b^c$, and obtain

$$\frac{\partial \zeta_C}{\partial t} = \frac{\nabla_{\|} \cdot \left(\rho_b^{mix} m_b^{c,eq} \int_{\zeta_C}^{\zeta_M} \widehat{u}_b dx_3 \right) - \Psi_{\mathcal{M}_b^c,\beta}\{\Xi\}}{\phi \rho_b^{mix} m_b^{c,eq}} + (1 - s_{b,R}) \frac{\partial \zeta_R}{\partial t} + (s_{b,R} - s_{b,T}) \frac{\partial \zeta_M}{\partial t}$$

$$\equiv \mathcal{D}_C \Xi.$$

Here we have proceeded as with the hysteresis model to define the evolution operator for ζ_C as $\mathcal{D}_C$. We can now give the analog of Equation (4.32) for the region of convective mixing as

$$\frac{\partial \zeta_C}{\partial t} = \begin{cases} \min\left(\mathcal{D}_C, \frac{\partial \zeta_R}{\partial t} \right) & \text{if} \quad \zeta_C = \zeta_R \\ \mathcal{D}_C & \text{if} \quad \zeta_B < \zeta_C < \zeta_R \\ \max(\mathcal{D}_C, 0) & \text{if} \quad \zeta_C = \zeta_B \end{cases}. \tag{4.33}$$

To apply our model of convective mixing, we need an understanding of what convective mixing looks like on the macroscale. Numerical experiments and stability analysis strongly indicate that convective mixing starts during the late meso temporal scale, after which the effective macroscale dissolution rate is nearly linear. For stationary sharp interfaces, good macroscale models for convective mixing are starting to appear in literature; however, there is still much to be done before a full characterization of $\Psi_{\mathcal{M}_b^c,\beta}\{\Xi\}$ is known. Further reading can be found in the references, and we will show some early applications of these ideas in Chapter 5 (see also Box 4.2).

4.9.4 The Complete Model

We have now provided three extra constraints on the macroscale independent variables, which we chose as the three heights $h_{T,M}$, $h_{M,R}$, and $h_{R,C}$ together with the coarse pressure for each phase. For completeness, we give the expressions for the coarse-scale mass in terms of the independent variables h:

$$M^c = \mathcal{M}_b^c + \phi \rho_c^{mix} m_c^{c,eq} \cdot \left[s_{c,R} h_{M,R} + s_{c,T} h_{T,M} \right]$$

and

$$M^b = \phi \rho_b^{pure} h_{C,B} + \phi \rho_b^{mix} m_b^{b,eq} \cdot \left[h_{R,C} + s_{b,R} h_{M,R} + s_{b,T} h_{T,M} \right] + \phi \rho_c^{mix} m_c^{b,eq} \cdot \left[s_{c,R} h_{M,R} + s_{c,T} h_{T,M} \right].$$

The complete model now consists of the two mass conservation equations (Eq. 4.6), as well as the two evolution equations (Eq. 4.32) and (Eq. 4.33). Finally, the model is closed with the pressure–interface relationship, which is given in the case of Dupuit reconstruction by Equation (4.31).

BOX 4.2 *Stability of Coarse Convective Mixing Model*

The core objective of multiscale modeling is to obtain models on different scales that describe the same underlying physical processes. As such, the properties of the models often share similar characteristics. In the case of convective mixing, this is a process that is of interest since it is unstable, and chaotic, on the fine scale. These instabilities develop with a characteristic wave length, which is referred to as the most unstable mode. Generation and propagation of instabilities can be characterized by a velocity that is proportional to $\rho_b^{mix} - \rho_b^{pure}$.

Our coarse-scale model as described in this section makes no effort to dampen the instabilities. As such, the coarse-scale model we have developed is indeed also unstable. This is most easily realized by considering a region where there is only brine, in other words $\zeta_T = \zeta_M = \zeta_R$. Our model now simplifies, and the equations are indeed identical to the simple sharp interface model described in Section 4.6. There, we saw that the coarse capillary pressure was given by Equation (4.25); however, the density difference for the present case has the opposite sign, since the denser fluid resides on top of the lighter fluid. We will see in Section 5.1.1 that the derivative of the (coarse) capillary pressure plays an important role, as it is the coefficient of the second-order, smoothing term. When the fluids are segregated with the heavier fluid on top, the derivative will have an opposite sign. Thus, instead of the coarse capillary pressure leading to a smoother solution, it will amplify variations: the solution is unstable.

Convective mixing is therefore a fascinating example of preserving the (unstable) characteristic of the fine-scale equations in the coarse-scale model. We will touch on this same topic when discussing multiscale numerical models in Section 5.3. Unfortunately, the instabilities are amplified in the coarse model, since there is no smallest, most unstable, wavelength when instabilities are driven by negative diffusion. Therefore, three different alternatives may be suggested when using the model in this section in practice:

1. The maximum growth rate of instabilities in the coarse model is determined by the horizontal scale of resolution of the model, since the instability comes from the second-order differential term in space. Let the horizontal length scale of application be denoted $\ell_{\|}$. Then, an upper bound on the second derivative of an interface ζ can be estimated by Taylor expansion as

$$\nabla_{\|}^2 \zeta \leq \frac{4h_{T,B}}{\ell_{\|}^2}.$$

The growth rate of instabilities on the coarse scale will be proportional to this term. The term $\Psi_{\mathcal{M}_b^c,\beta}\{\Xi\}$ represents dissolution due to instabilities on all scales smaller than $\ell_{\|}$. Since we know from analysis of the fine-scale model that the growth of instabilities decays quickly with wave length, we expect there to exist a length scale above which instabilities on the coarse scale grow at a rate that is insignificant compared to the fine-scale instabilities. We express this as a condition on $\ell_{\|}$, and require

$$\rho_b^{mix} m_b^{c,eq} K \lambda_b (g \cdot e_3)(\rho_b^{pure} - \rho_b^{mix}) \frac{4h_{T,B}}{\ell_{\|}^2} \ll \Psi_{\mathcal{M}_b^c,\beta}\{\Xi\}.$$

The coefficients are approximated from an equation for saturation analogous to the procedure in Section 5.1.1, and K represents the characteristic horizontal coarse (integrated) permeability.

If a numerical approximation is used, then $\ell_{\parallel}$ corresponds to the typical horizontal dimension of the grid, and will for practical applications usually satisfy the constraint.

2. If a finer grid than indicated above is desired (or an analytical solution), then dominant convective fingers will be resolved by the grid. The limiting characteristic of finger growth is the resistance to vertical flow. If the impact of vertical flow is not accounted for in the pressure reconstruction, we will therefore overestimate finger growth in the coarse model. When spatially fine grids are desired in the coarse model, we must therefore resort to using vertically structured pressure reconstructions such as the one given in Section 4.7. The additional complexity introduced in the coarse model by using a more accurate pressure reconstruction has yet to be fully investigated.
3. Finally, the unstable term is proportional to the brine density difference $\left(\rho_b^{mix} - \rho_b^{pure}\right)$, which is usually on the order of only a few kilograms per cubic meter. While it is this difference that drives instabilities on the fine scale, one may consider to model this density difference as insignificant on the coarse-scale. This would avoid all problems with an unstable coarse scale model which now relies exclusively on the term $\Psi_{\mathcal{M}_b^c,\beta}\{\Xi\}$ to model the impact of convective mixing.

4.10 DISCUSSION OF COARSE MODELS

In this chapter we have constructed a suite of different coarse-scale models. They are characterized not only by differences in complexity, but also by differences in modeling assumptions, embodied in a variety of reconstructions associated with different physical processes being modeled. We summarize these models in this section, in terms of the main ideas in the reconstruction.

4.10.1 Pressure Reconstruction

We have considered two options to reconstruct pressure. The simplest is the Dupuit reconstruction. This is based on the assumption of negligible fluid flow in the direction normal to the top and bottom boundaries, which is also the direction of upscaling, identified by unit vector $\boldsymbol{e}_3$. This implies that the fluid pressure in the x_3-direction can be reconstructed from fluid statics. The simplicity of this approach has allowed us to combine it with all of our reconstructions for other variables. As such, the Dupuit assumption and associated reconstruction may be seen as a fundamental building block for the methods presented in this chapter. The alternative to the Dupuit reconstruction is the vertically structured pressure reconstruction, based on the idea of linear variation in fluid velocities in the $\boldsymbol{e}_3$ direction. This concept is easiest to realize in the context of sharp interfaces, but even then, it is relatively complex. In general, the additional complexity of this reconstruction as compared to the Dupuit reconstruction makes it more restrictive in its application. However, the increased quality of reconstruction available with the vertical structure can be a valuable tool, in particular in near-well regions where vertical flows are not negligible, as we will see in Chapter 5 when discussing interface upconing near wells.

4.10.2 Saturation Reconstruction

The saturation reconstruction has been based on a combination of the Dupuit reconstruction for pressure and the fine-scale relationship between capillary pressure and saturation, constrained by the assumption of buoyant segregation of the fluids. This, in general, gives what we have referred to as the capillary fringe model, with saturation changing continuously along the x_3-direction. When the capillary fringe can be ignored, it is often convenient to replace the continuous variation of saturation with a sharp interface description, where the saturation changes abruptly across a given surface. The usual sharp interface model assigns only one mobile fluid to any region along the x_3-line, which leads to significant simplification of the model. We considered a model with regions of mobile CO_2 (with residual brine saturation), residual CO_2 (with mobile brine), and brine at full saturation. Due to its simplicity, the sharp interface approach has a versatility that allows it to be combined with all our other reconstruction models. In contrast, the capillary fringe model is more complex and is thus more difficult to combine with other reconstructions. However, it provides an increased accuracy that may be important, as we will see in the next chapter when discussing self-similar solutions to injection problems.

4.10.3 Extended Reconstructions

We have complemented the basic building blocks of pressure and saturation reconstruction with more advanced models. In particular, we have presented two reconstructions that address specific physical complexities. First, we have considered how to extend the reconstructions to account for changes in fluid densities. Second, we considered the important problem of phase transitions with a focus on dissolution of CO_2. We addressed this problem by introducing tertiary, model-dependent variables, with a concomitant extra equation to model the transfer of CO_2 components from the CO_2 phase to the brine phase.

By close inspection, these models reveal how physical phenomena may lead to terms in the equations with different mathematical structure on different scales. The clearest example is the gravitational force, which acts as a body force in the microscale equations. Yet, on the macroscale, we have seen how the gravity force acts not only as a body force, but also enters our equations in the form of a coarse-scale capillary pressure. In contrast, the capillary pressure on the microscale introduces no terms on the macroscale, but its effect can clearly be seen within the saturation reconstruction in the fact that our mobilities become nonlinear, where they would have been linear in the absence of fine-scale capillary pressure. A final example of change of character appears in the equations for mass transfer. On the microscale, we modeled all phase partitioning as instantaneous. When we develop our coarse-scale model, however, the mass transfer is no longer instantaneous, and its coefficients may become time-dependent.

These changes in mathematical character may perhaps not be surprising to some readers; however, the importance of these changes cannot be overstated. In particular, when we design numerical methods, our numerical system will be significantly different if we discretize the macro- or microscales, precisely because the physical processes involved have changed their mathematical character.

4.11 FURTHER READING

Book

LAKE, L. 1996. *Enhanced Oil Recovery*. Englewood Cliffs, NJ: Prentice Hall.

Papers

E, W. and B. ENGQUIST. 2003. The heterogeneous multi-scale methods. *Communications in Mathematical Sciences*, 1, 87–133.

GASDA, S. E., J. M. NORDBOTTEN, and M. A. CELIA. 2011. Vertically averaged approaches for CO_2 migration with solubility trapping. *Water Resources Research*, 47, W05528, doi: 10.1029/2010WR009075.

NORDBOTTEN, J. M. and H. K. DAHLE. Impact of the capillary fringe in vertically integrated models for CO_2 storage. *Water Resources Research*, 47, W02537, doi: 10.1029/2009WR008958.

YORTSOS, Y.C. 1955. A theoretical analysis of vertical flow equilibrium. *Transport in Porous Media*, 18, 107–129.

Chapter 5

Solution Approaches

In this chapter, we will discuss solution approaches to the vertically integrated, coarse-scale governing equations developed in Chapter 4. To some extent, this material will generalize and build upon the solutions for well hydraulics, as well as the numerical methods, developed in Chapters 2 and 3. In Section 5.1 we present analytical solutions, as well as some example calculations, for near-well problems involving both CO_2 injection wells and leaky passive wells. In Section 5.2 we frame traditional numerical methods in the language of the multiscale modeling developed in Chapter 4. We demonstrate how the concepts of compression and reconstruction become tools not only for transitions between scales and dimensions, but also between continuous and discrete equations. With this perspective, the introduction of multiscale numerical methods in Section 5.3 becomes a natural extension.

The final section of this chapter is devoted to showing how these solutions can be used to model systems involving real-world data. We have chosen three examples. In the first example, we use the analytical solutions for near-well flow to analyze the spatial impact of a typical CO_2 storage operation. The second example considers flow under a sloping caprock on the temporal megascale. Finally, the third example analyzes a specific proposed storage formation in the North Sea using numerical methods. Together, these three examples show the applicability of analytical and numerical solutions associated with the vertically integrated, coarse-scale governing equations. They also introduce some additional understanding of the length and time scales faced in real applications. We will return to this theme again in Chapter 6.

5.1 ANALYTICAL SOLUTIONS

The full equations for CO_2 storage in three dimensions are too complex, both in terms of parameters and processes, to readily admit analytical solutions. In contrast, several of the vertically integrated macroscale models presented in the previous chapter have a sufficiently simple structure that analytical solutions can be obtained for relevant problems. These solutions serve three purposes. By themselves, analytical solutions provide insight into the interplay among different processes and parameters in the

Geological Storage of CO_2: Modeling Approaches for Large-Scale Simulation, First Edition.
Jan M. Nordbotten and Michael A. Celia.

problem. Together with numerical approximations, analytical solutions are valuable from the perspective of verification and benchmarking. Finally, analytical solutions serve as building blocks for fast algorithms presented in the next chapter.

In this section we present two approximate analytical solutions for two-phase flow problems. The first is an analysis of CO_2 injection into an initially brine-filled aquifer, exploiting self-similarity. The second is a discussion of models for interface upconing around a pumping (or leaking) well.

5.1.1 Injection into a Confined Formation

Injection into a confined aquifer defines the first major time period of a CO_2 storage operation. During this period, advective two-phase flow dominates the system, while the dip of the aquifer, dissolution, and mineral reactions are expected to have minimal influence on the solution. Therefore, we simplify the system by assuming a horizontal aquifer with the immiscible incompressible model of Section 4.5. We also use the Dupuit pressure reconstruction from Section 4.4. We model the injection operation on the temporal mesoscale and the spatial macroscale.

We now consider injection into a confined homogeneous and isotropic aquifer with constant fluid properties, where the injection takes place at a constant rate through a single vertical well. If we choose a flat pressure datum ζ_p, with $x_3 = z$, the problem has radial symmetry and we write the coarse-scale phase equations in radial coordinates as

$$\Phi \frac{\partial S_\alpha}{\partial t} - \frac{1}{r}\frac{\partial}{\partial r}\left(r\Lambda_\alpha(S_\alpha)K\frac{\partial}{\partial r}P_\alpha \right) = 0, \quad (\alpha = c, b). \tag{5.1}$$

The initial and boundary conditions for this problem are

$$S_c(t=0) = 0.$$
$$S_c(r=\infty) = 0.$$
$$\lim_{r\to 0} -2\pi r\Lambda_\alpha(S_\alpha)K\frac{\partial}{\partial r}P_\alpha = Q_\alpha.$$

Here Q_α are the volumetric injection rates. Note that both Q_c and Q_b must be non-negative in order to ignore fluid extraction.

Development of Self-Similar Solutions

We can write the radial fractional flow form of Equation (5.1) by eliminating the pressures. This is achieved by summing the phase equations, integrating over r with application of the inner boundary condition, relating the phase pressures to one another using the capillary pressure, and rearranging, to obtain,

$$\Phi \frac{\partial S_c}{\partial t} + \frac{1}{r}\frac{\partial}{\partial r}\left(\frac{\Lambda_c(S_c)}{\Lambda_\Sigma(S_\alpha)}\frac{Q_\Sigma}{2\pi} - rK\frac{\Lambda_\Pi(S_\alpha)}{\Lambda_\Sigma(S_\alpha)}\frac{\partial P^{cap}(S_c)}{\partial r} \right) = 0, \tag{5.2}$$

where subscript Σ denotes summation over phases while subscript Π denotes multiplication. Physically, the problem has a natural scaling introduced by the total injected volume, $V(t) = tQ_\Sigma$. If we consider a cylindrical distribution of the injected fluid, the volume would be expressed as $V(t) = \pi r^2\Phi$. Elimination of $V(t)$ provides a dimensionless group linking space and time:

$$\chi \equiv \frac{\pi\Phi}{Q_\Sigma}\frac{r^2}{t}.$$

Note that this definition of the dimensionless group χ differs slightly from the definition applied by the authors in the references, which included a factor of 2 and incorporated the concept of drainable porosity. It is also different from the dimensionless group that appeared in Section 2.4. If we assume we can write the solution as $S_c(\chi)$, we can express the derivatives by using the chain rule as follows:

$$\frac{\partial}{\partial t} = \frac{\partial \chi}{\partial t}\frac{d}{d\chi} = -\frac{\chi}{t}\frac{d}{d\chi} \quad \text{and} \quad \frac{\partial}{\partial r} = \frac{\partial \chi}{\partial r}\frac{d}{d\chi} = 2\frac{\chi}{r}\frac{d}{d\chi}.$$

After substitution into Equation (5.2), we see that the solution must satisfy the following dimensionless equation:

$$-\chi\frac{dS_c}{d\chi} + \frac{d}{d\chi}\left(\frac{\Lambda_c(S_c)}{\Lambda_\Sigma(S_\alpha)} - 2\Gamma\chi\frac{\Lambda_\Pi(S_\alpha)}{\Lambda_b'\Lambda_\Sigma(S_\alpha)}\frac{d}{d\chi}\Pi^{cap}(S_c)\right) = 0, \tag{5.3}$$

where we have defined a dimensionless capillary pressure as suggested by Equation (4.26), by normalizing by the difference in hydrostatic pressures over the thickness of the aquifer

$$\Pi^{cap}(S_c) \equiv \frac{P^{cap}(S_c)}{\Delta_\alpha \rho g h_{T,B}} = \Pi_c - \Pi_b.$$

Note that the normalized phase pressures have the same scaling. We use the convention for primes on mobilities $\Lambda_\alpha' \equiv \Lambda_\alpha(S_\alpha = 1)$, and we have defined the ratio of gravitational to advective forces as follows, where the coefficient of 2 is included for historic reasons:

$$\Gamma = 2\frac{\pi\Lambda_b' K \Delta_\alpha \rho g h_{T,B}}{Q_\Sigma}.$$

Thus, the self-similar solution $S_c(\chi)$ satisfies an ordinary differential equation (ODE), as given in Equation (5.3), with the second-order term depending on the coarse-scale capillary pressure. Note that the saturation depends on the single dimensionless variable χ, rather than the original variables r and t. This is an important observation, because it implies that we only need to integrate a single ODE to know the solution of the injection problem for all space and time. The integration may be performed analytically or numerically. In the following, we first present the general numerical solution, then derive two analytical solutions under different sets of assumptions.

The numerical solution to Equation (5.3) involves a slight complication: the equation is defined when $\chi > 0$, and also $0 < S_c < 1$. At the two end points of the interval, Equation (5.3) degenerates to a first-order equation. This has two implications. First, it implies that the tip has finite propagation speed, so that there exists some point χ_0 such that $S_c(\chi_0) = 0$. Second, because we have a first-order equation at this point, we can interpret this as an expression for the first derivative of the solution,

$$\left[\frac{1}{\Lambda_b'}\left(1-2\Gamma\chi_0\frac{d\Pi^{cap}}{dS_c}\frac{dS_c}{d\chi}\right)\frac{d\Lambda_c}{dS_c}-\chi_0\right]\frac{dS_c}{d\chi}=0.$$

We see that this equation has two solutions. If $S_c/d\chi = 0$, then the solution is constant and equal to S_c. This will be the case for $\chi > \chi_0$. Otherwise, the derivative is given by the solution to the expression in the square brackets, which implies

$$\lim_{\chi\uparrow\chi_0}\frac{dS_c}{d\chi}=\frac{1-\chi_0\Lambda_b'\left(\dfrac{d\Lambda_c}{dS_c}\right)^{-1}}{2\Gamma\chi_0\dfrac{d\Pi^{cap}}{dS_c}}. \tag{5.4}$$

Note that since the derivative of S_c must be negative at χ_0, Equation (5.4) provides a lower bound on the outer tip of the plume given by

$$\chi_0 > \frac{1}{\Lambda_b'}\frac{d\Lambda_c}{dS_c}.$$

Solution of Equation (5.3) involves integration from χ_0 inward, using Equation (5.4) at the end point, and continuing with any standard numerical method for Equation (5.3) when $S_c > 0$. However, since the outer point χ_0 is not known, an iterative search procedure must be implemented using the constraint on χ_0 that the solution satisfies the correct volume balance:

$$Q_c=\frac{d}{dt}\int_0^\infty 2\pi r\Phi S_c(\chi(r))\,dr = Q_\Sigma\frac{d}{dt}t\int_0^\infty S_c(\chi(r))\,d\chi,$$

which implies that the integral of the solution must be the volume fraction of CO_2 injected,

$$\int_0^{\chi_0} S_c(\chi(r))\,d\chi=\frac{Q_c}{Q_\Sigma}. \tag{5.5}$$

We note that for injection of CO_2 this fraction is one; for mixed injection of CO_2 and brine it will be less than one.

We also note that the coefficient multiplying the dimensionless capillary pressure term in Equation (5.3) scales as $\sim Q_\Sigma^{-1}$, so that in the limit of high injection rates, we may omit the second-order term from Equation (5.3). This greatly simplifies the equation, allowing us to derive an analytical solution. With the capillary pressure term eliminated from the equation, we can evaluate the remaining derivative by parts to obtain

$$\left[\frac{d}{dS_c}\left(\frac{\Lambda_c(S_c)}{\Lambda_\Sigma(S_\alpha)}\right)-\chi\right]\frac{dS_c}{d\chi}=0.$$

This equation shows that the solution is either constant, or must satisfy the expression in the brackets,

$$\chi(S_c) = \frac{d}{dS_c}\left(\frac{\Lambda_c(S_c)}{\Lambda_\Sigma(S_\alpha)}\right). \tag{5.6}$$

This is an expression for the self-similar coordinate in terms of the solution, which can be inverted to find $S_c(\chi)$. We note that the right-hand side is the derivative of the coarse flux function, and thus in self-similar coordinates, the solution is analogous to that for the Buckley–Leverett equation in Cartesian coordinates, with the only difference being the scaling induced by the self-similar variable. As with the Buckley–Leverett equation, the solution to Equation (5.6) may be multi-valued. The correct limiting solution is then obtained by replacing the expression in parenthesis in Equation (5.6) by its convex envelope, as reviewed in Section 3.6.

We may further simplify the system using the sharp interface model of Section 4.6. This is of particular interest in the high-injection-rate case (which corresponds to $\Gamma << 1$), and for early time, since the local flow pattern is often too fast for a significant capillary fringe to form. When the mobilities in Equation (5.6) are obtained from a sharp interface model, we recall that they will be linear functions of S_c. We also recall that the initial condition is brine saturated, so that $s_{c,B} = 1 - S_{b,B} = 0$, and observe that when the saturation behind the interface is at the residual saturation of brine, $\lambda_b(s_{b,T}) = 0$. Then from Section 4.6 we know that

$$\Lambda_c(S_c) = \frac{S_c}{s_{c,T}} h_{T,B}\lambda_c(s_{c,T}), \quad \text{and} \quad \Lambda_b(S_c) = \left(1 - \frac{S_c}{s_{c,T}}\right) h_{T,B}\lambda_b(s_{b,B}).$$

With these linear coarse-scale mobilities, we can invert Equation (5.6) explicitly for $S_c(\chi)$:

$$\frac{S_c(\chi)}{s_{c,T}} = \frac{h_{T,M}}{h_{T,B}} = \begin{cases} 0 & \text{if} \quad s_{c,T}\chi \geq \lambda \\ \dfrac{1}{\lambda - 1}\left(\sqrt{\dfrac{\lambda}{s_{c,T}\chi}} - 1\right) & \text{if} \quad \lambda^{-1} < s_{c,T}\chi < \lambda \\ 1 & \text{if} \quad 0 < s_{c,T}\chi \leq \lambda^{-1} \end{cases} \tag{5.7}$$

Recall that λ is the fine-scale mobility ratio. The integral from $\chi = 0$ to $\chi = \lambda/s_{c,T}$ of the solution $S_c(\chi)$ defined in Equation (5.7) is 1; thus, from the volume-balance constraint given in Equation (5.5), we see that Equation (5.7) solves a problem with pure CO_2 injection. The analogous solution to the sharp interface problem with fractional injection, that is, co-injection of CO_2 and brine, is

$$\frac{S_c(\chi)}{s_{c,T}} = \frac{h_{T,M}}{h_{T,B}} = \begin{cases} 0 & \text{if} \quad s_{c,T}\chi \geq \lambda \\ \dfrac{1}{\lambda - 1}\left(\sqrt{\dfrac{\lambda}{\chi s_{c,T}}} - 1\right) & \text{if} \quad s_{c,T}\chi_* < s_{c,T}\chi < \lambda \\ \left(1 + \dfrac{Q_b}{Q_c}\lambda\right)^{-1} & \text{if} \quad 0 < \chi \leq \chi_* \end{cases} \tag{5.8}$$

The point χ_* is defined as the point where the expressions in Equation (5.8) intersect, that is,

$$\chi_* = \frac{\lambda}{s_{c,T}\left(\frac{Q_c(\lambda-1)}{Q_c+Q_b\lambda}+1\right)^2}.$$

The part of the solution given in Equation (5.8) that is defined by $S_c(\chi)/s_{c,T} = (1+(Q_b/Q_c)\lambda)^{-1}$ satisfies the governing ODE, since $dS_c/d\chi = 0$, and can be interpreted as giving the correct asymptotic ratio of flow rates into the well. It also guarantees that Equation (5.8) satisfies the volume balance constraint. However, we also note that Equation (5.8) is not valid when $Q_b < 0$, since it leads to coarse-scale saturations (or interface) outside the physical range. This is a feature of self-similar solutions—it is not always possible to construct a self-similar solution satisfying the boundary conditions, even though the equation itself may indicate a self-similar structure.

We will often need to consider the pressure profile, which can be obtained by eliminating $\partial S_b/\partial t$ from Equations (5.1) and (5.2) and integrating the resulting equation to find:

$$P_b - P_{b,0} = -\int_R^r \frac{\Lambda_c(S_c)}{\Lambda_\Sigma(S_\alpha)}\frac{dP^{cap}(S_c)}{dS_c}\frac{\partial S_c}{\partial r} + \frac{Q_\Sigma}{2\pi r'\Lambda_\Sigma(S_\alpha)K}dr'.$$

Here R represents an outer boundary where $P_b = P_{b,0}$. If $S_c(R) = 0$, we recognize the first term of the integral from Equation (3.25) and Equation (4.26) as the difference between the CO_2 pressure and the global pressure, so that the global pressure satisfies

$$P_G - P_{G,0} = \frac{Q_\Sigma}{2\pi K}\int_R^r \frac{dr'}{r'\Lambda_\Sigma(S_\alpha)} = \frac{Q_\Sigma}{4\pi K}\int_{\chi_R}^{\chi} \frac{d\chi'}{\chi'\Lambda_\Sigma(S_\alpha)}. \tag{5.9}$$

The simple form of Equation (5.9) emphasizes the utility of the global pressure concept. In particular, we see that the global pressure is linearly proportional to the injection rate, a fact that will be very useful in Chapter 6. Further, once the global pressure is known, it is easy to construct the coarse phase pressures that drive the fluid flow. For the special case of high injection rate, where Equation (5.8) is applicable, Equation (5.9) can be integrated algebraically to yield the dimensionless coarse global pressure (in terms of χ, and assuming $\chi_R > \lambda/s_{c,T}$):

$$\Gamma\Pi_G(\chi) = \begin{cases} \Gamma\Pi_{G,0} + 1/2\ \ln \chi_R/\chi & \text{if} \quad s_{c,T}\chi \ge \lambda \\ \Gamma\Pi_G(\lambda/s_{c,T}) + 1 - \sqrt{\frac{s_{c,T}\chi}{\lambda}} & \text{if} \quad s_{c,T}\chi_* \le s_{c,T}\chi < \lambda. \\ \Gamma\Pi_G(\chi_*) + \frac{\Lambda_b'}{2\Lambda_\Sigma(S_c(\chi_*))}\ln \chi_*/\chi & \text{if} \quad 0 < \chi < \chi_* \end{cases} \tag{5.10}$$

The fraction in the last line reduces to the inverse mobility ratio λ^{-1} when we are considering pure CO_2 injection. Further, we have written the equation in terms of the dimensionless group $\Gamma \Pi_G$, which stays bounded even though the physical pressure becomes unbounded in the high-injection-rate limit. This is natural, since the required pressure drive increases as the injection rate increases. In our experience, the closed form expressions obtained for high injection rates are a very good approximation when $\Gamma < 0.1$, after which the approximation properties worsen. For $\Gamma > 1$, it is generally advisable to solve the full ODE given in Equation (5.3). In general, more care must be taken when using the closed form expression for pressure (Eq. 5.10), than the expression for saturation (Eq. 5.8), due to the pressure becoming unbounded in the high-injection-rate limit.

As a final comment, we note that it is not always obvious that self-similar solutions are stable solutions for a problem. For the case explored in this section, stability can be shown at the limit of high injection rates, and numerical experiments indicate stability also for moderate and low injection rates.

Example Solutions

The simplicity of the self-similar analysis in this section motivates us to use the problem of injection into a confined aquifer to highlight the role of different parameters in our coarse-scale model. In particular, we assess the impact of capillary forces, the mobility ratio, the injection rate, and the presence of an injected water phase on the spread of CO_2 into the aquifer.

To avoid considering the full nonlinear variability of the capillary pressure and relative permeability curves, we simply consider typical functional forms, and investigate the impact of scaling their magnitude. By choosing the same nondimensionalization of fine-scale pressure as we used for the coarse-scale equation, we give the fine-scale capillary pressure curve as

$$\frac{p^{cap}}{\Delta_\alpha \rho g h_{T,B}} = c_{cap} \cdot (1 - s_{c,N})^{-1/2},$$

where we have introduced the saturations normalized by the irreducible brine saturation, $s_{c,N} \equiv s_c/(1 - s_{b,res})$ and $s_{b,N} = 1 - S_{c,N}$. Further, the nondimensional constant c_{cap} scales strength of the capillary forces relative to the pressure difference over a hydrostatic column spanning the aquifer. As a matter of choice, we consider cubic relative permeability functions, such that

$$\lambda_c(s_{c,N}) = s_{c,N}^3 \lambda_c' \quad \text{and} \quad \lambda_b(s_{c,N}) = (1 - s_{c,N})^3 \lambda_b'.$$

We will not consider the coefficients (end-point values) λ_c' and λ_b' individually, since we know from the analysis that only their ratio matters, which we recall is denoted as $\lambda = \lambda_c'/\lambda_b'$. The self-similar solution can now be evaluated once the dimensionless gravity number Γ (also interpreted as the inverse injection rate) and the CO_2 fraction Q_c/Q_Σ have been fixed.

For typical parameters ($\Gamma = 1$; $Q_c/Q_\Sigma = 1$; $\lambda = 5$; and $c_{cap} = 0.1$), we can solve Equation (5.3) numerically, and visualize the reconstructed saturation profile as

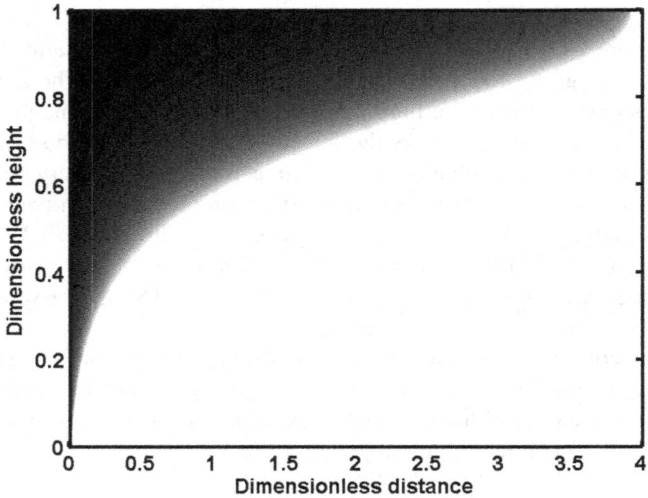

Figure 5.1 Fine-scale saturation reconstruction for the case with capillary fringe. Note that the saturation is normalized.

shown in Figure 5.1. From the reconstructed saturation, we see that the capillary fringe spans about 20% of the thickness of the formation. Since the extent of the fringe is directly proportional to c_{cap}, we may already conjecture capillary forces to be dominant when $c_{cap} > 1$, and a sharp interface approximation to be appropriate when $c_{cap} < 0.01$. Figure 5.1 also allows us to appreciate that near the tip of the CO_2 plume, CO_2 will exist at lower saturations than closer to the well. Thus, the mobility of CO_2 at the plume tip will be lower, which leads to a slight buildup of CO_2 at the tip, rather than a thin tip which tapers off.

We now proceed to investigate the impact of the four different dimensionless groups separately, fixing the parameters given above as our base case.

Gravity Number (Inverse Injection Rate) We consider a range of values for the gravity number, with $\Gamma = \{0, 0.1, 1, 10, 100, 1000\}$. For $\Gamma = 0$, we use Equation (5.6), which is shown with a solid line in Figure 5.2. Note how the saturation profile develops a discontinuity, since there is no longer a second-order term in the equation to smooth the solution. For nonzero values of the gravity number, we again solve Equation (5.3) numerically, and display the results with dashed lines in Figure 5.2. These results support a notion that Equation (5.6) may be a suitable approximation to Equation (5.3) when $\Gamma < 0.1$; however, for larger values of the gravity number, we see that gravitational forces start to dominate, leading to a much flatter saturation profile. From a storage efficiency perspective, this argues for the natural strategy of injecting at as high a rate as permissible.

Mobility Ratio The range of physical mobility ratios is much smaller than the range of possible gravity numbers. As such, we only consider values of $\lambda = \{1, 2,$

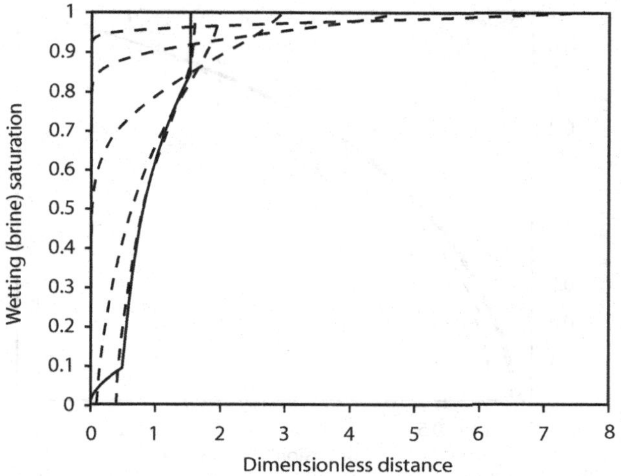

Figure 5.2 Impact of gravity number on injection profiles. Solid line is asymptotic limit as $\Gamma < 0.1$.

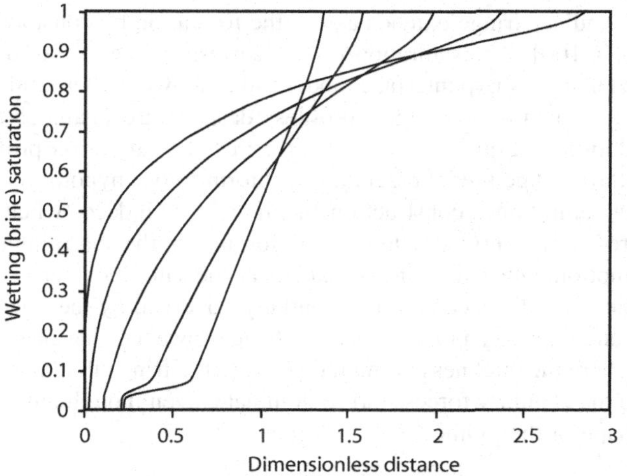

Figure 5.3 Impact of mobility ratio on injection profiles, with steeper lines corresponding to lower values.

5, 10, 20}. The results are shown in Figure 5.3, where the steepest lines correspond to the lowest mobility ratios. We see that the mobility ratio is important for the sweep efficiency of the CO_2 plume, where low mobility ratios lead to an almost piston-like displacement. Such observations suggest engineering solutions wherein the brine residual saturation is manipulated such that the mobility ratio is minimized.

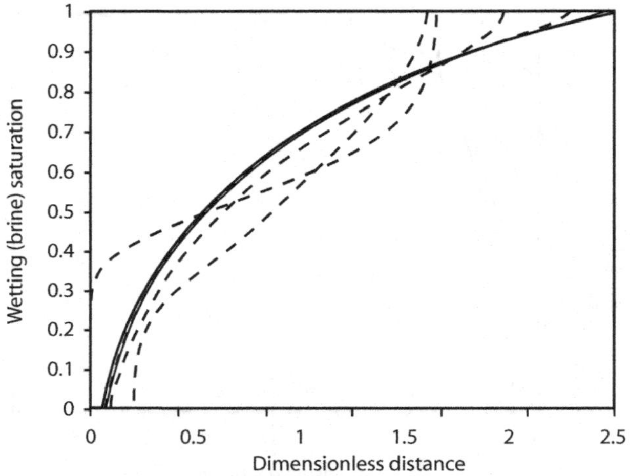

Figure 5.4 Impact of capillary fringe on injection profiles. Solid line corresponds to no capillary fringe (sharp interface).

Capillary Fringe We explore the range of models from a sharp interface model to where the capillary fringe extends across the formation by considering $c_{cap} = \{0, 0.01, 0.1, 1, 10, 100\}$. The saturation profile corresponding to the lowest value of c_{cap} is obtained using a sharp interface model and is shown with a solid line in Figure 5.4, while the remaining saturation profiles (dashed) are again obtained from a numerical solution of Equation (5.3). We see that the saturation profiles obtained with the capillary fringe model essentially conform to our hypothesis after visually inspecting the saturation reconstruction in Figure 5.1. Indeed, when the fringe is small compared to the vertical extent of the formation, the sharp interface is a reasonable assumption, while for fringes that smear the saturation across the thickness of the formation, the displacement is essentially governed by the fine-scale relative permeability and capillary pressure curves. Note, however, that near the tip of the CO_2 plume, where the thickness is small relative to the fringe thickness, there comes a point where the capillary forces start to dominate, even though they do not impact the overall shape of the plume.

Mixed Injection Finally, we explore the possibility of mixed injection of CO_2 and brine, considering volume fractions of $Q_c/Q_\Sigma = \{0.1, 0.25, 0.5, 0.75, 1\}$. Within models of vertically equilibrated saturation distributions, the results illustrated in Figure 5.5 show that such strategies uniformly reduce the volume of aquifer rock exposed to CO_2, and thus reduce the potential for residual trapping. This also exemplifies the importance of the length and time scales. Indeed, at time scales before gravity segregation has taken place, mixed injection of CO_2 and brine can potentially be optimized to create exactly the effect of lowered mobility ratio suggested above. The optimal strategy therefore depends not only on the properties of the problem, but also on the time scale over which the solution is considered.

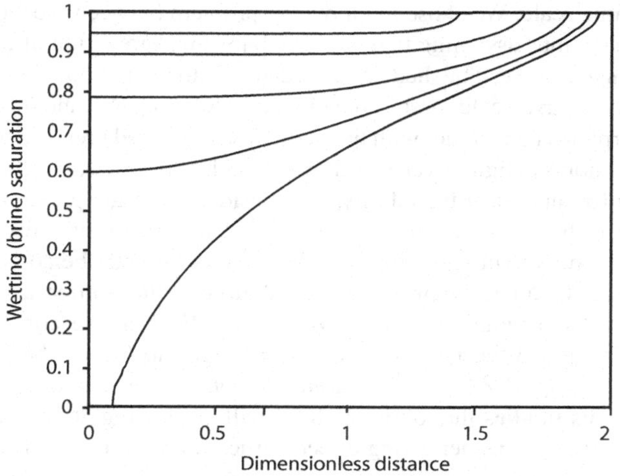

Figure 5.5 Impact of mixed injection of brine and CO_2 on injection profiles.

5.1.2 Interface Upconing Near a Well

Fluid flow to a well is an important problem with a long history in porous media flow research. We have already seen examples in Chapter 2 of solutions to single-phase flow problems appearing in groundwater hydrology. Of equal historical importance are problems where there is a dense fluid under a lighter fluid, such as the case when an aquifer contains a layer of fresh water above a layer of saltwater, or when recovering petroleum resources where some combination of gas, oil, and water is present. In terms of CO_2 storage, this problem is again central, since leakage through a well will appear locally as an extraction well from a layer of CO_2, with the CO_2 segregated above a layer of brine (if we are above the temporal microscale). However, there is an important difference between the historical problems and the current: for both groundwater and petroleum extraction, the flow (extraction) rates in the well can be controlled. For both problems, it is beneficial to avoid pulling the internal fluid–fluid interface into the well, since the lower fluid is frequently not desired. Therefore, the well is operated so that the interface is minimally perturbed, and analytical solutions can exploit this, because the goal is to assess the largest extraction rate that maintains a small perturbation of the interface. In contrast, leakage of CO_2 through an abandoned well is not a controlled process, and the flow rate may be high enough that the fluid–fluid interface is strongly perturbed. Indeed, high leakage rates correspond to the cases of greatest interest. Therefore, many of the traditional approximate solutions to this problem are outside their range of validity for the parameters encountered when modeling CO_2 storage.

When modeling upconing, we are interested in a problem that is on the spatial meso- and macroscales. The temporal scale is more nuanced. The well leakage itself is on the microscale, yet we expect the movement of the fluids in the aquifer to be on a greater time scale. We chose to model the problem between the temporal meso- and macroscales, thus assuming that we have buoyant segregation of fluids, but that a capillary fringe is not established. In accordance with the notion of temporal scales, we will use the coarse-scale models developed in Sections 4.6 and 4.7, corresponding to the sharp interface model with the vertically structured pressure reconstruction to account for non-negligible vertical flow near a leaking well.

To allow for analytical tractability, we consider, as in the previous section, the aquifer to be isotropic, homogeneous, and horizontal in the vicinity of the leaky well, and we model the fluids as immiscible. We choose both the coordinate system and the pressure datum as the pressure at the bottom of the aquifer, $\zeta_P = \zeta_B = 0$. The equations then have radial symmetry with $x_3 = z$. Further, we consider the brine to be immobile above the interface, $\lambda_b(s_{b,T}) = 0$, and no CO_2 to be present below the interface, $0 = s_{c,B} = \lambda_c(s_{c,B})$. To simplify notation, we denote $\lambda_c = \lambda_c(s_{c,T})$ and $\lambda_b = \lambda_b(s_{b,B})$. This implies that only one fluid will be flowing at any point in space, and we only need to consider a single reconstructed pressure, which we define as

$$\hat{p} = \begin{cases} \widehat{p_b} & \text{if} \quad z \le \zeta_M \\ \widehat{p_c} & \text{if} \quad z > \zeta_M \end{cases}.$$

We are primarily interested in the steady-state interface upconing, and we consider the time-independent form of Equation (5.1) in radial coordinates, written in terms of the fine-scale flux with no sources or sinks:

$$\frac{d}{dr}\left(r \int_{\zeta_B}^{\zeta_T} u_{\alpha,r} dx_3 \right) = 0.$$

From this it follows that

$$\int_{\zeta_B}^{\zeta_T} u_{\alpha,r} dx_3 = \frac{Q_\alpha}{2\pi r}. \tag{5.11}$$

Keep in mind that for flow to a well, the flow rates Q_α will be negative. The radial flux $u_{\alpha,r}$ is approximated using reconstructions given as

$$u_{\alpha,r} \approx \widehat{u_{\alpha,r}} = \mathcal{R}_{u_{\alpha,r}}^{VS} \Xi = -\lambda_\alpha \left(\mathcal{R}_{s_\alpha}^{SI} \Xi \right) k_r \frac{d}{dr} \left(\mathcal{R}_p^{VS} \Xi \right). \tag{5.12}$$

Note that subscript r denotes radial direction. Also note that Equation (5.12) has no hysteresis because the problem is time-independent. We can write an explicit expression for the vertically structured (VS) pressure reconstruction by evaluating the integrals in Equation (4.30) along with the application of the interface condition of Equation (4.29). Then, for steady-state conditions, the vertical flow component at the interface is related to the radial component. The integrated expression for pressure then takes the following form,

$$\hat{p} = \mathcal{R}_p^{VS} \Xi = \begin{cases} P - \rho_b g z - \dfrac{z^2}{2h_{M,B}} \dfrac{u_{b,r}^-}{\lambda_b' k_z} \dfrac{d\zeta_M}{dr} & \text{if} \quad z \le \zeta_M \\ \hat{p}(\zeta_M) - \rho_c g (z - \zeta_M) - \dfrac{h_{T,M}^2 - (\zeta_T - z)^2}{2h_{T,M}} \dfrac{u_{c,r}^+}{\lambda_c' k_z} \dfrac{d\zeta_M}{dr} & \text{if} \quad z > \zeta_M \end{cases}. \tag{5.13}$$

Note that pressure reconstruction contains the unknown fluid velocities $u_{b,r}^{-}$ and $u_{c,r}^{+}$ at the interface. We can at any time simplify Equation (5.13) to the Dupuit reconstruction of pressure by neglecting the nonlinear terms, which is equivalent to considering the limit of $k_z \to \infty$. The system is considered on a domain $r \in (0, R]$, and constrained by boundary conditions of the type

$$h_{M,B}(r = R) = h_{M,B}^{R}$$
$$P(r = R) = P^{R}$$
$$\lim_{r \to 0} -2\pi r \int_{\zeta_B}^{\zeta_T} \widehat{u_{\alpha,r}} dx_3 = Q_\alpha$$

where superscript R denotes evaluation at the outer boundary, $r = R$. To simplify later expressions, we will of course also take $h_{T,M}^{R} = h_{T,B} - h_{M,B}^{R}$ as a known boundary condition. We will now review four different ways to simplify the coupled equations describing the upconing problem.

The Muskat Solution

Muskat was one of the first researchers to address the problem of interface upconing. He used the two coarse-scale phase pressures as independent variables (as we discussed in Section 4.4), using the outer boundary condition as an initial approximation to ζ_M, and the Dupuit approximation to derive a boundary condition for the second pressure:

$$\Delta_\alpha P^{R} = g \Delta_\alpha \rho h_{M,B}^{R}.$$

Then, by assuming that $u_{\alpha,r}$ had negligible vertical variation, he applied Equations (5.11) and (5.12) to approximate the pressure in the lighter fluid (in our case, the CO_2 pressure) as

$$P_c \approx P_c^{R} - \frac{Q_c}{2\pi \Lambda_c \left(h_{T,M}^{R}\right) K_r} \ln \frac{r}{R}.$$

Similarly, for the brine pressure we have

$$P_b \approx P_b^{R} - \frac{Q_b}{2\pi \Lambda_b \left(h_{M,B}^{R}\right) K_r} \ln \frac{r}{R}.$$

We recall from Section 4.5 that when the macroscale system has the two phase pressures as independent variables, we can obtain saturation (or equivalently the interface) as a dependent variable. Applying the Dupuit reconstruction, we get the coarse-scale capillary function given in Equation (4.25), from which we obtain the interface as approximated by Muskat:

$$h_{M,B} = \frac{\Delta_\alpha P}{\Delta_\alpha \rho g} = h_{M,B}^{R} + \left(\frac{Q_c}{\Lambda_c \left(h_{T,M}^{R}\right)} - \frac{Q_b}{\Lambda_b \left(h_{M,B}^{R}\right)} \right) \frac{\ln r/R}{2\pi K_r \Delta_\alpha \rho g}. \tag{5.14}$$

This solution has the property that it is a function of ln r only; thus, the Muskat approximation will always lead to a straight line on a semilog plot.

Dupuit Upconing Solution

A more advanced approach to approximating the solution to the upconing problem is to consider the feedback between the interface location and the pressure profile. For this case, we will use one coarse pressure and the interface location as independent variables. We begin by considering the flow of brine subject to the Dupuit assumption, which can be expressed in integral form as follows,

$$\int_{\zeta_B}^{\zeta_M} \widehat{u_{b,r}}dx_3 = -\Lambda_b(h_{M,B})K_r \frac{d}{dr}(\mathcal{R}_p^D P) = -\Lambda_b(h_{M,B})K_r \frac{dP}{dr} = \frac{Q_b}{2\pi r}.$$

Similarly, for the flow of CO_2 we write,

$$\int_{\zeta_M}^{\zeta_T} \widehat{u_{c,r}}dx_3 = -\Lambda_c(h_{T,M})K_r \frac{d}{dr}(P - \Delta_\alpha \rho g h_{M,B}) = \frac{Q_c}{2\pi r}.$$

We can eliminate the pressure derivative between these two equations, to obtain

$$2\pi r K_r \Delta_\alpha \rho g \frac{dh_{M,B}}{dr} = \frac{Q_c}{\Lambda_c(h_{T,M})} - \frac{Q_b}{\Lambda_b(h_{M,B})}. \tag{5.15}$$

This equation is separable, and can be integrated, yielding a transcendental equation, as we will see in the next section.

Vertically Structured Solution

The approach to solve the full system, using the vertically structured reconstruction, is similar to that of the Dupuit upconing solution. However, due to the increased nonlinear couplings, the equations cannot be solved exactly. Eventually, it is necessary to approximate the horizontal fluxes appearing in equation (5.13) using an assumption of essentially uniform magnitude (as would follow from a Dupuit approximation). As such, we make the following approximations for the flows of brine and CO_2, motivated by steady-state flow conditions and the flow rates at the well,

$$u_{b,r}^- \approx \frac{Q_b}{2\pi r h_{M,B}} \quad \text{and} \quad u_{c,r}^+ \approx \frac{Q_c}{2\pi r h_{T,M}}.$$

With these approximations, we can derive the extension of Equation (5.13), which takes the form of a nonlinear second-order differential equation. The solution is well approximated by the lowest order terms, which we write as

$$-\left(2\pi r K_r \Delta_\alpha \rho g h_{T,B}^{-1} - \frac{k_r}{k_z}\frac{Q_c\lambda_b' + Q_b\lambda_c'}{3r\lambda_b'\lambda_c'}\right)\frac{dh_{T,M}}{dr} = \frac{Q_c}{\lambda_c' h_{T,M}} - \frac{Q_b}{\lambda_b' h_{M,B}}. \tag{5.16}$$

Here we have used that for homogeneous sharp interface models, the coarse mobilities are linear and can be expressed as $\Lambda_c(h_{T,M}) = \lambda_c' h_{T,M} h_{T,B}^{-1}$ and $\Lambda_b(h_{M,B}) = \lambda_b' h_{M,B} h_{T,B}^{-1}$.

Equation (5.16) can again be integrated to yield a transcendental equation. Let the constants in the first and second terms on the left-hand side be denoted c_1 and

c_2, the constants on the right-hand side be denoted c_3 and c_4, and write the equation in terms of $h_{T,M}$:

$$-\left(c_1 r - \frac{c_2}{r}\right)\frac{dh_{T,M}}{dr} = \frac{c_3}{h_{T,M}} - \frac{c_4}{h_{T,B} - h_{T,M}}.$$

The solution to this differential equation must satisfy

$$\int\left(c_1 r - \frac{c_2}{r}\right)^{-1} dr + \int\left(\frac{c_3}{h_{T,M}} - \frac{c_4}{h_{T,B} - h_{T,M}}\right)^{-1} dh_{T,M} = constant.$$

Evaluating the integrals (and collecting terms), we see that the left-hand side is given by

$$J\left(h_{T,M}, r\right) \equiv \frac{1}{2c_1}\ln\left(c_1 r^2 - c_2\right) + \frac{h_{T,M}^2}{2(c_3 + c_4)} - \frac{c_4 h_{T,M} h_{T,B}}{\left(c_3 + c_4\right)^2}$$
$$- \frac{c_3 c_4 h_{T,B}^2}{\left(c_3 + c_4\right)^2}\ln\left(c_4 h_{T,M} - c_3\left(h_{T,B} - h_{T,M}\right)\right).$$

Using the boundary conditions to determine the constant value, we see that the solution $h_{T,M}(r)$ must satisfy

$$J\left(h_{T,M}, r\right) = J\left(h_{T,M}^R, R\right). \tag{5.17}$$

Note that since c_2 is the only constant that depends on k_z, and $c_2 \to 0$ when $k_z \to \infty$, we recover the solution of the Dupuit upconing model from the previous section by setting $c_2 = 0$ in the expression for J.

An interesting observation can be made from the integrated form J of either Equation (5.15) or Equation (5.16), which is that as $R \to \infty$, the first logarithm term also tends to $+\infty$. Thus, for the expression $J(\zeta_T - \zeta_{M,R}, R)$ to stay bounded, the second logarithm term must tend to $-\infty$. Since the coefficient is positive (remember that Q_α are negative for pumping), this will be the case if the argument tends to zero, in other words, we have that

$$\lim_{R\to\infty} \frac{Q_b}{\Lambda_b\left(h_{M,B}\right)} - \frac{Q_c}{\Lambda_c\left(h_{T,M}\right)} = 0. \tag{5.18}$$

This has important implications in practice, if the equation is interpreted as a constraint on Q_c/Q_Σ in terms of the outer boundary condition. Equation (5.18) can then be used to determine the smallest possible volumetric fraction of CO_2 which can flow through the well,

$$\frac{Q_c}{Q_\Sigma} > \frac{\Lambda_c\left(h_{T,M}^R\right)}{\Lambda_b\left(h_{M,B}^R\right)}\frac{Q_b}{Q_\Sigma} = \frac{\Lambda_c}{\Lambda_\Sigma}.$$

Thus, the flow of CO_2 up through a well will always exceed what is predicted by the fractional flow formulation, which is to be expected since CO_2 is buoyant compared to water.

Upconing with a Capillary Fringe

We look at a final case, which goes slightly beyond the scope we laid out in the beginning of this section. If we are interested in modeling upconing where a capillary fringe might be important, such as slow leakage occurring over longer time scales, we can include this in the derivation of a Dupuit upconing model. Note, however, that the vertically structured pressure reconstruction is not defined other than for sharp interface type models; therefore, we do not attempt to simultaneously capture vertical structure in the flow field and a capillary fringe. The derivation is identical to that presented for the Dupuit upconing model, and can also be seen directly by considering the steady state of Equation (5.2), leading to the equation

$$2\pi rK\frac{dP^{cap}(S_c)}{dS_c}\frac{dS_c}{dr}=\frac{Q_c}{\Lambda_c(S_c)}-\frac{Q_b}{\Lambda_b(S_b)}. \tag{5.19}$$

Note that this equation is not always analytically integrable. In cases where it is not, care needs to be taken when attempting to integrate Equation (5.19), as it is highly unstable when integrating inward from the outer boundary condition. In cases where integrating outward from a known inner boundary condition is of interest, Equation (5.19) poses no problems.

5.2 NUMERICAL METHODS

The goal of numerical simulation is to compute sufficiently accurate solutions to practical problems of interest. This challenge is met through a combination of appropriate grid design, accurate numerical methods, and computational algorithms that allow for solution of the resulting equations within reasonable time. While for many applications accurate solutions can be calculated, this is not necessarily the case for CO_2 storage, and particularly not for fine-scale three-dimensional models. Here, the size of the computational domain in space and time, together with the coupled non-linear processes acting on a large spectrum of scales, produces computational challenges that have yet to be overcome.

Our primary interest lies in the coarse-scale equations described in the previous sections. These coarse-scale equations are in many cases more linear than the original fine-scale equations (recall the behavior of the relative permeability functions). Further, our coarse-scale equations are defined in only two spatial dimensions. Thus, there is hope that the computational challenge can, to some extent, be met for these coarse models.

Numerical methods for problems in porous media are the subject of rich and interesting research. It is well beyond the scope of this section to give an overview that justifies their importance. Instead, we will give a short, modern exposition of numerical methods in the framework of the compression and reconstruction operators with which we have become familiar (see Figure 5.6), emphasizing the common links between numerical methods and the multiscale approaches of Chapter 4. In this context, the numerical representation will be the coarse space, while the

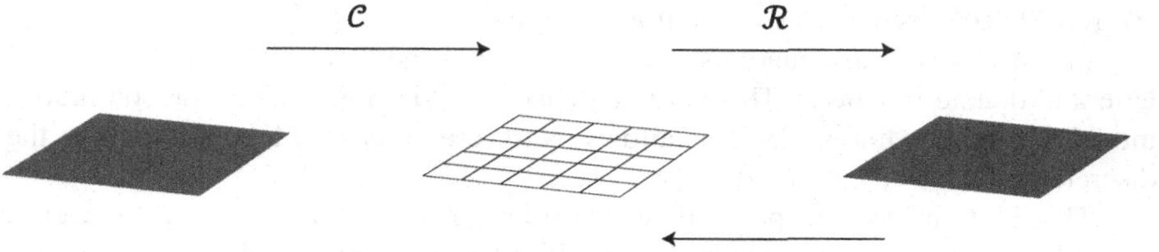

Figure 5.6 Conceptual sketch of multiscale modeling of continuous and discrete domains.

continuous problem defines the fine space, even though the continuous problem may be on the macro- or megascales. The multiscale formulation of numerical methods will also prepare the ground for the review of upscaling and grid-based multiscale methods of the next section.

Our model equations in this section are based on the global pressure/fractional flow formulation of the immiscible constant-density equations from Section 4.6, and we consider a horizontal domain such that the coarse-scale gravity does not enter. Further, we do not consider tensor coarse-scale mobilities here. The derivation of the model equations given below is identical to the exposition of the fractional flow formulation in Chapter 3. For the coarse-scale equations, the global pressure equation takes the form of an elliptic equation (see Eqs. (3.15) and (3.24)):

$$-\nabla_{\|}\cdot(\Lambda_{\Sigma}\boldsymbol{K}\nabla_{\|}P_G)=Q_{\Sigma}.$$

The saturation is governed by the following mass balance equation for the CO_2 phase,

$$\Phi\frac{\partial S_c}{\partial t}+\nabla_{\|}\cdot\boldsymbol{U}_c=Q_c.$$

The phase flux is given by the fractional flow formulation,

$$\boldsymbol{U}_c=\frac{\Lambda_c}{\Lambda_{\Sigma}}\boldsymbol{U}_{\Sigma}-\frac{\Lambda_{\Pi}}{\lambda_{\Sigma}}\boldsymbol{K}\nabla_{\|}P^{cap}\equiv\boldsymbol{F}(S_c),$$

where the total flux is given in terms of the global pressure

$$\boldsymbol{U}_{\Sigma}=-\Lambda_{\Sigma}\boldsymbol{K}\nabla_{\|}P_G.$$

We refer to $\boldsymbol{F}(S_c)$ as the flux function. From these equations, we see that we need to define discrete approximations to elliptic equations (which we refer to as the pressure equation) of the form

$$-\nabla\cdot(\boldsymbol{k}\nabla p)=q \tag{5.20}$$

and to hyperbolic (when the gradient term in $\boldsymbol{F}(S_c)$ is negligible) and parabolic (otherwise) equations of the form

$$\frac{\partial s}{\partial t}+\nabla\cdot\boldsymbol{f}(s)=0. \tag{5.21}$$

We refer to this equation as the saturation equation.

In these two final equations, we have omitted the $\parallel$ subscript, and use small letters to denote functions. This is to emphasize that in this section, the continuous model will be our fine scale. This allows us to reserve capital letters to denote the discrete model.

This section has four parts. First, we briefly review what is meant by a grid. Second, we review some methods for discretizing elliptic problems. Then, we discuss temporal discretizations. Finally, we discuss discretizations in space and time. These sections generalize the earlier introductions to numerical methods in Chapters 2 and 3.

5.2.1 The Grid

The basis for numerical discretizations is in most cases a grid, which can be thought of as a coarse representation of our original domain (see Figure 5.6). In this section, we will define notation for a general *unstructured* grid. Unlike the *structured* grids used in Chapter 2, these unstructured grids consist of arbitrary polygons, although frequently some regularity is required to ensure that specific discretizations lead to good approximations.

We consider the grid as a nonoverlapping division of the domain Ω into a finite number of smaller domains ω_i, termed grid cells, which we number by $i = 1 \ldots N_C$, where N_C is the number of cells. We will only be interested in grids composed of polygonal cells, and we subdivide cell boundaries into line segments $\partial\omega_{i,j}$, termed cell edges, where $j = 1 \ldots n_{E,i}$, where $n_{E,i}$ is the number of edges of cell ω_i. We will also need to be able to reference the edges simply as $\partial\omega_i$, where $i = 1 \ldots N_E$, and N_E refers to the total number of edges in the grid. By $\boldsymbol{\nu}_{i,j}$ we denote the normal vector to an edge $\partial\omega_{i,j}$, pointing outward with respect to cell ω_i, and we denote by $\boldsymbol{\nu}_i$ the normal vector to an edge $\partial\omega_i$, pointing according to an established convention (what convention is used is not important, as long as it is consistent). For each cell ω_i we identify a point $\boldsymbol{x}_i \in \omega_i$, which will usually be the midpoint of the cell. The vertexes of the cells are identified as $\boldsymbol{y}_{i,j}$, where $j = 1 \ldots n_{V,i}$, and $n_{V,i}$ is the number of vertexes of cell ω_i. We will also need to be able to reference the vertexes simply as y_i, where $i = 1 \ldots N_V$, where N_V is the total number of vertexes in the grid.

Numerical methods can be designed in terms of cells, the cell edges, or the cell vertexes. For conservation laws, it is often most appropriate to use the cells, and this is what we will emphasize in the following.

5.2.2 Pressure Equation

In this section, we will use the multiscale framework to present spatial discretizations to variational forms of the elliptic Equation (5.20). To obtain a variational formulation, we multiply the original problem with a suitably smooth function, called a test function, and then integrate by parts. This moves the derivatives to the test function, whose properties are known. If this new problem has a solution that is independent of the choice of test functions, then, loosely speaking, the solution to the variational problem is the same as the solution to the original problem.

Variational Formulation

In the context of our elliptic model problem, we commonly choose test functions w that are once differentiable (at least almost everywhere), and zero at the boundary. Let us denote the set (space) of all such functions as $\underline{w}$. If p solves the original problem, then for all test functions w in the space of test function $\underline{w}$, it also holds that

$$-w\nabla \cdot (\boldsymbol{k}\nabla p) = wq.$$

Moreover, this equation also holds if we integrate over our domain Ω. We define the notation

$$(a, b) \equiv \int_{\Omega} a \cdot b d\boldsymbol{x}$$

for arbitrary functions a and b. Using this notation with integration by parts and the zero values of w along the boundary $\partial\Omega$, we have that

$$-(\nabla \cdot (\boldsymbol{k}\nabla p), w) = (\boldsymbol{k}\nabla p, \nabla w) = (q, w). \tag{5.22}$$

This equation is meaningful, subject to the integrals involved being finite. Theoretically, this is of importance, but it will not play a major role in this section, and we will omit the technical details.

The Finite Element Method

Finite element methods are designed to solve variational problems such as Equation (5.22) by representing the functions p and w by a linear combination of a finite set of functions. These methods usually focus on unknowns at the vertexes of the grid. Let us assume that we know that the solution p can be found within some space of functions $\underline{p}$. Here, and in the following, notation with an underscore denotes a function space. Finite approximations of $\underline{p}$ and $\underline{w}$ are coarser representations (the original spaces are infinite-dimensional), and are usually defined on a grid. Let us for the moment assume that our grid consists of triangles. The simplest finite approximation to (almost everywhere) differentiable functions is then the space of functions that are linear (planar in two dimensions) on each element, and continuous across the edges. This finite approximation space, denoted $\underline{p_h}$, is a subspace of the original space, $\underline{p_h} \subset \underline{p}$, and the subscript h emphasizes the dependence on the grid, where h is usually interpreted to represent a characteristic length scale for the grid size. An illustration of a function $p \in \underline{p}$, and an approximation to it, $p_h \in \underline{p_h}$, is shown in Figure 5.7 for a one-dimensional system. Also shown in Figure 5.7 are two piecewise linear basis functions, along with an indication of the grid size, h. Note how the approximation properties are dependent not only on the grid, but also on the regularity of the solution: visually we can confirm that $p_h(x)$ may be a reasonably good approximation over the left-hand side of the figure, but if the space $\underline{p_h}$

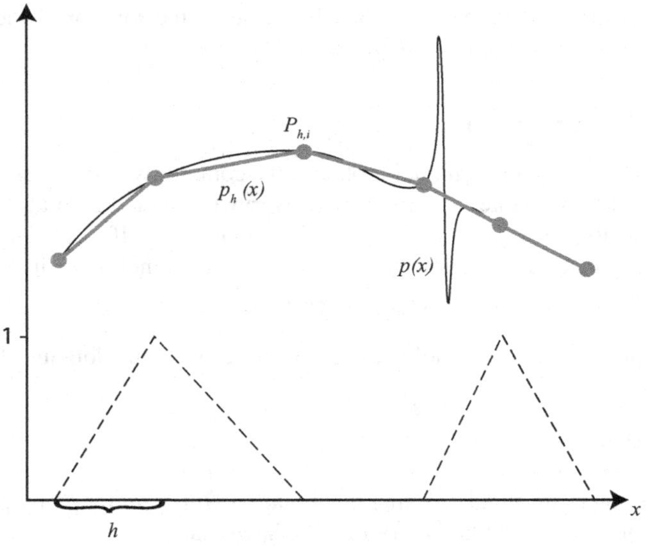

Figure 5.7 Finite elements applied in 1D.

is too coarse for $p_h(x)$ to capture the oscillation in $p(x)$ illustrated in the right-hand side of the figure.

Basis functions in this example are functions $\vartheta \in \underline{p_h}$ that are chosen as piecewise linear polynomials such that they take the value $\vartheta_i(\boldsymbol{y}_i) = 1$ at one vertex of the grid, and the value $\vartheta_i(\boldsymbol{y}_j) = 0$ for all other vertexes of the grid, $i \neq j$. The basis functions form a complete basis for $\underline{p_h}$. In other words, every function $p_h \in \underline{p_h}$ can be uniquely defined by a linear combination of the basis functions ϑ_i. Given the nodal values of the basis functions (i.e., values at the vertexes), the expansion for p_h may be written as follows,

$$p_h(\boldsymbol{x}) = \sum_{i=1}^{N_V} p_h(\boldsymbol{y}_i)\vartheta_i(\boldsymbol{x}). \tag{5.23}$$

The node values of p_h are our coarse representation and form the unknowns in the system to be solved. We refer to them as $P_{h,i}$, and to the vector composed of all node values as $\boldsymbol{P}_h = [P_{h,1} \ldots P_{h,N_V}]^T$ (see again Figure 5.7 for an illustration).

We can express this notion in terms of the compression operators we have become familiar with,

$$P_{h,i} = \mathcal{C}_P^{FE} p \equiv p(\boldsymbol{y}_i).$$

Here the superscript *FE* designates a finite element compression, which is simply sampling the continuous (fine-scale) function. The reconstruction operator is defined by Equation (5.23), such that we have

$$p_h = \mathcal{R}_p^{FE} \boldsymbol{P}_h \equiv \sum_{i=1}^{N_V} P_{h,i} \vartheta_i.$$

Note that the subscript h plays the same role in denoting a reconstructed function as the hat did in Chapter 4. We choose to use subscript h in this section to honor the traditional notation, where h also emphasizes the dependence of the solution on the grid resolution. A finite representation $\underline{w_h}$ of $\underline{w}$ is usually obtained in the same manner, although in general the spaces or basis functions may be different. Therefore, we denote the basis functions of the test space by φ_i, even though it is common to take $\varphi_i = \vartheta_i$. The test functions can be expressed in terms of the reconstruction as $w_h = \mathcal{R}_w^{FE} \boldsymbol{a}$, where $\boldsymbol{a}$ is a vector of rank N_V. The finite element method for Equation (5.22) can now be expressed as: find the coarse solution vector $\boldsymbol{P}_h$ such that

$$\left(k \nabla \mathcal{R}_p^{FE} \boldsymbol{P}_h, \nabla \mathcal{R}_w^{FE} \boldsymbol{a}\right) = \left(q, \mathcal{R}_w^{FE} \boldsymbol{a}\right) \tag{5.24}$$

holds for all rank N_V vectors $\boldsymbol{a}$. Since the problem is linear, it is sufficient to consider all vectors $\boldsymbol{e}_i$, for which we have that $\mathcal{R}_w^{FE} \boldsymbol{e}_i = \varphi_i$. Substituting the finite basis representations into Equation (5.24), we then obtain the coarse problem: find the solution vector $\boldsymbol{P}_h$ such that

$$\sum_{j=1}^{N_V} P_{h,j} \left(\boldsymbol{k} \nabla \vartheta_j, \nabla \varphi_i\right) = \left(q, \varphi_i\right)$$

holds for all i. Since all the integrals in this problem are independent of the solution, we can evaluate them a priori to obtain the matrix $\boldsymbol{A}$ and the vector $\boldsymbol{q}$ whose entries are defined by

$$A_{i,j} = \left(\boldsymbol{k} \nabla \vartheta_j, \nabla \varphi_i\right) \quad \text{and} \quad q_i = \left(q, \varphi_i\right).$$

This allows us to write the finite element problem as a matrix-vector problem,

$$\boldsymbol{A} \boldsymbol{P}_h = \boldsymbol{q}.$$

Furthermore, the matrix $\boldsymbol{A}$ is sparse, since for most pairs of basis functions φ_i and ϑ_j, there is no overlap, and hence the integral defining the entry $A_{i,j}$ is zero.

Unfortunately, for the problems we are interested in, this simple finite element method does not have ideal properties. Most notably, while the solution is usually a good approximation to pressure, the fluxes are often not accurate. This becomes a particular concern when the pressure equation is coupled to a saturation equation, or when the flow properties are highly heterogeneous. Further, since the compression and reconstruction operators are not conservative, the finite element method may introduce mass conservation errors when approximating conserved quantities, such as mass (or saturation for incompressible problems). Using more accurate approximation spaces within the finite element context generally does not alter this evaluation.

Mixed Variational Problem

Our pressure equation is derived from a conservation law together with a flux expression. These two equations together represent the physics of the problem. It is therefore natural to think of numerical methods that preserve this structure of the equations. We start by considering the continuous problem in which the balance equation and the flux equation are written separately as follows,

$$\nabla \cdot \boldsymbol{u} = q \quad \text{and} \quad \boldsymbol{u} = -\boldsymbol{k}\nabla p. \tag{5.25}$$

In the same way as for the standard variational problem, we can now multiply each of these equations by a test function, noting that the second equation is a vector equation. After integrating over the domain and by parts, and assuming that the boundary conditions are such that all boundary terms are zero, we then arrive at the following formulation,

$$\begin{aligned} (\nabla \cdot \boldsymbol{u}, w) &= (q, w) \\ (\boldsymbol{k}^{-1}\boldsymbol{u}, v) - (p, \nabla \cdot v) &= 0. \end{aligned} \tag{5.26}$$

The test functions $w \in \underline{w}$ and $v \in \underline{v}$ are of course in different spaces, and this gives rise to the name "mixed" variational problem. The natural constraint for Equation (5.26) to be meaningful is that the potential p and test function w should be square integrable, while both the flux and divergence of $\boldsymbol{u}$ and v must also be square integrable. Note that this is a less restrictive assumption on the potential than was needed for the finite element method.

Mixed Finite Elements

If the finite spaces for the solutions $\boldsymbol{u}$ and p, and for the test functions v and w, are chosen to be the same, Equation (5.26) retains its symmetric nature, and the resulting discrete linear system will be symmetric. This is almost always the choice, and we will therefore assume that both $\boldsymbol{u} \in \underline{v}$ and $p \in \underline{w}$. Still, the finite approximation to the vector space $\underline{v}$ and the scalar space $\underline{w}$ cannot be chosen independently, they must be compatible. The compatibility condition is known as the inf-sup condition. The lowest order approximation that satisfies the inf-sup condition for our model problem is given in terms of the Raviart–Thomas finite elements, denoted by RT0. For an RT0 approximation, the scalar spaces $\underline{w_h}$ are chosen as constant on each element ω_i, while the vector spaces $\underline{\boldsymbol{v}_h}$ are chosen such that elements $v_h \in \underline{\boldsymbol{v}_h}$ are constant over each edge $\partial\omega_{i,j}$, with the appropriate extrapolation to the internal part such that $\nabla \cdot v_h \subset \underline{w_h}$, and $\nabla \times v_h = \boldsymbol{0}$ internal to each element.

The basis functions for the RT0 scalar space can be defined as

$$\vartheta_i(\boldsymbol{x}) = \begin{cases} 1 & \text{if} \quad \boldsymbol{x} \in \omega_i \\ 0 & \quad otherwise \end{cases}.$$

Then the vector space basis function $\boldsymbol{\vartheta}_j \in \underline{\boldsymbol{v}_h}$ can be defined such that

$$\boldsymbol{\vartheta}_{j_1} \cdot \boldsymbol{V}_{j_2} = \begin{cases} 1 & \text{if} \quad j_1 = j_2 \\ 0 & \quad otherwise \end{cases}.$$

Thus, the discrete scalar space $\underline{w_h}$ has the same dimension as the number of cells N_C, while the discrete vector space $\underline{\boldsymbol{v}_h}$ has the same dimension as the number of edges N_E. We define compression operators from the original spaces to the coarse finite spaces such that the scalar compression takes the mean value over the grid block,

$$P_{h,i} = \mathcal{C}_P^{MFE} p \equiv \frac{(p, \vartheta_i)}{(\vartheta_i, \vartheta_i)}.$$

Similarly, the vector compression takes the mean value over the edge,

$$U_{h,i} = \mathcal{C}_U^{MFE} \boldsymbol{u} \equiv \frac{\int_{\partial\omega_i} \boldsymbol{u} \cdot \boldsymbol{\vartheta}_i dx}{\int_{\partial\omega_i} \boldsymbol{\vartheta}_i \cdot \boldsymbol{\vartheta}_i dx}.$$

We see that these compression operators are conservative when the coarse variables are interpreted as integrated values, contrary to the sampling compression used in the regular finite element method. The solution vectors are defined as before, with $\boldsymbol{P}_h = [P_{h,1} \ldots P_{h,N_C}]^T$ and $\boldsymbol{U}_h = [U_{h,1} \ldots U_{h,N_E}]^T$. We obtain the reconstruction operators from the finite element basis functions,

$$p_h = \mathcal{R}_p^{MFE} \boldsymbol{P}_h \equiv \sum_i P_{h,i}\vartheta_i \quad \text{and} \quad \boldsymbol{u}_h = \mathcal{R}_u^{MFE} \boldsymbol{U}_h \equiv \sum_i U_{h,i}\boldsymbol{\vartheta}_i.$$

With these tools in hand, we define the mixed finite element approximation as: find the solution vectors $\boldsymbol{P}_h$ and $\boldsymbol{U}_h$ such that

$$\begin{aligned} \left(\nabla \cdot \mathcal{R}_u^{MFE} \boldsymbol{U}_h, \mathcal{R}_p^{MFE} \boldsymbol{e}_i\right) &= (q, \mathcal{R}_p^{MFE} \boldsymbol{e}_i) \\ \left(\boldsymbol{k}^{-1} \mathcal{R}_u^{MFE} \boldsymbol{U}_h, \mathcal{R}_u^{MFE} \boldsymbol{e}_j\right) - \left(\mathcal{R}_p^{MFE} \boldsymbol{P}_h, \nabla \cdot \mathcal{R}_u^{MFE} \boldsymbol{e}_j\right) &= 0 \end{aligned} \tag{5.27}$$

for all $i = 1 \ldots N_C$ and all $j = 1 \ldots N_E$. As with the standard finite elements, this system defines a linear set of equations, where the system matrix can be computed a priori and has a sparse structure.

The mixed finite element formulation presented in Equation (5.27) has advantages over the finite element formulation in Equation (5.24). Most notably, the mixed formulation explicitly calculates fluxes, which are of primary interest, and ensures that these fluxes have physical meaning. Consequently, not only are the compression operators for the mixed finite element method consistent with conservation, but the whole discrete system explicitly represents the conservation. This comes at the cost of more unknowns, and also a linear system that is no longer positive definite.

Control Volume Methods

Recall that in the section on finite element methods, we left open the possibility of choosing a different approximation space for the test functions than for the solution. Then, in the section on mixed finite element methods, we saw that scalar test functions that are piecewise constant on each element could be of interest, since the associated compression is consistent with conservation. If we combine these ideas, it leads us to a new derivation of the control volume methods presented in Chapter 2.

We define the test functions as in the mixed finite element section, using piecewise constants,

$$\varphi_i(\boldsymbol{x}) = \begin{cases} 1 & \text{if} \quad \boldsymbol{x} \in \omega_i \\ 0 & \quad \textit{otherwise.} \end{cases}$$

Using these basis functions in the finite element framework leads to

$$(q, \varphi_i) = (\boldsymbol{k}\nabla \mathcal{R}_p \boldsymbol{P}_h, \nabla \varphi_i) = -\sum_{j=1}^{n_{E,i}} \int_{\partial \omega_{i,j}} \boldsymbol{v}_{i,j} \cdot (\boldsymbol{k}\nabla \mathcal{R}_p \boldsymbol{P}_h)\, dx,$$

which has to hold for all $i = 1 \ldots N_C$. We refer to

$$u_{h,i,j} \equiv -\int_{\partial \omega_{i,j}} \boldsymbol{v}_{i,j} \cdot \boldsymbol{k}\nabla \mathcal{R}_p \boldsymbol{P}_h dx$$

as cell edge fluxes, and we see that we can define numerical methods based on constructing approximations to $u_{h,i,j} = u_{h,i,j}(\boldsymbol{P}_h)$ from $\boldsymbol{P}_h$:

$$(q, \varphi_i) = \sum_{j=1}^{n_{E,i}} u_{h,i,j}(\boldsymbol{P}_h). \tag{5.28}$$

An important observation is that Equation (5.28) is actually the original form of the conservation law, where the left-hand side is the integral of sources and sinks, while the right-hand side is the integral of the flux exiting and entering the volume. As such, control volume methods are usually (as was our approach in Section 2.6) defined from Equation (5.28) directly, without the connection to variational methods.

The obvious way to approximate $u_{h,i,j}$ is by choosing some reconstruction operator $\mathcal{R}_p$. The simplest choice, using the piecewise linear finite elements introduced earlier, $\mathcal{R}_p = \mathcal{R}_p^{FE}$, requires a dual grid, since we are interested in fluxes crossing cell boundaries. This leads to what is known as the control volume finite element method. Alternatively, one can obtain approximations to $u_{h,i,j}$ without explicitly reconstructing pressure by either using a simple finite difference across the interface (as described in Chapter 2), or by more elaborate means.

The advantage of control volume methods is that they have no more degrees of freedom than the regular finite element formulation, yet they approximate the flux directly, thus in some way forming a compromise between mixed finite element methods and finite element methods.

5.2.3 Saturation Equation

The second equation governing our model problem of flow in porous media is the saturation equation. This equation is characterized by its strong locality, its space-time coupling, and the nonlinearities of the flux function. In this section we give a brief overview of some of the standard approaches to solving this equation.

As with the pressure equation, we start by introducing a class of test functions; however, since this is a problem in space-time, our test functions will also be in space-time. After integrating over space-time and then by parts, we obtain from Equation (5.21)

$$\left[s, \frac{\partial w}{\partial t}\right] + [f(s), \nabla w] = 0. \tag{5.29}$$

Here we have used square brackets to indicate space-time integration,

$$[a, b] \equiv \int_0^\infty (a, b)dt = \int_0^\infty \int_\Omega a \cdot b dx dt.$$

Because of the causality of the problem with respect to time dimension, we need not solve for the whole time domain at once, but we can move forward in time step-by-step. This is a great computational advantage. We therefore make the assumption that the spatial-temporal grid has a structure using discrete times $t_0, t_1, \ldots t_{N_T}$, such that no spatial-temporal grid cell crosses a time level t_k.

From the variational form of the hyperbolic problem given in Equation (5.29), we can define a control volume formulation analogously to that given in Section 5.2.2. In particular, if we assume that the spatial component of our grid is independent of time, we can denote the spatial-temporal grid by its spatial component and the time level it begins at, $\omega = \omega_i^{t_k}$. By again choosing our test function to be piecewise constant on the spatial-temporal grid, Equation (5.29) evaluates for $\varphi_i^{t_k}$ as

$$\int_{\omega_i} s(t_{k+1})dx - \int_{\omega_i} s(t_k)dx + \int_{t_k}^{t_{k+1}} \sum_{j=1}^{n_{E,i}} \int_{\omega_{i,j}} \mathbf{v}_{i,j} \cdot f(s) dx dt = 0. \tag{5.30}$$

From the first term, we see that it is natural to choose the integral (or average) saturation as a coarse variable, and we define the compression

$$S_{h,i}^{t_k} = C_s^H s(t_k) = \frac{1}{|\omega_i|} \int_{\omega_i} s(t_k) dx.$$

Here we have used the superscript H to indicate compression for a hyperbolic conservation law, and denoted the area of a grid cell i by $|\omega_i|$.

The simplest approach to define the saturation reconstruction is based on piecewise constants, for example,

$$s_h(x, t_k \le t < t_{k+1}) = \mathcal{R}_s^{H,ex} S_h^{t_k} \equiv \sum_i S_{h,i}^{t_k} \varphi_i(x). \tag{5.31}$$

With this reconstruction, our discrete version of Equation (5.30) is

$$S_{h,i}^{t_{k+1}} = S_{h,i}^{t_k} - \frac{1}{|\omega_i|} \int_{t_k}^{t_{k+1}} \sum_j \int_{\omega_{i,j}} \mathbf{v}_{i,j} \cdot f(s_h) dx dt. \tag{5.32}$$

With small enough time steps, the locality of the problem allows us to approximate the normal flux separately at each interface, using the original Equation (5.21) to propagate the initial condition $\mathcal{R}_s^{H,ex} S_h^t$. Since the reconstructed solution is piecewise constant, the initial condition for the local problem is a set of Riemann problems, which for one-dimensional scalar hyperbolic problems are relatively easy to solve. This approach is generally referred to as a Godunov method.

It turns out that the accuracy of Godunov's method is limited by the accuracy of the reconstruction, which in our example was lowest order. Improvements can be sought by using higher order reconstructions, and this is the basis for high-order methods for hyperbolic problems. Here, the challenge is to avoid oscillations that may be introduced by the higher order reconstructions.

The reconstruction given in Equation (5.31) leads to what is known as an explicit scheme (explaining the superscript *ex* on the reconstruction operator), where the fluxes are calculated based on the previous time step. The time steps that are permissible in order for explicit schemes to be stable are often prohibitively short. Therefore, implicit schemes are often desirable, as we mentioned in Section 3.8. The implicit scheme can be obtained in this setting by using a reconstruction based on the final time step,

$$s_h\left(\boldsymbol{x}, t_k \le t < t_{k+1}\right) = \mathcal{R}_s^{H,im} \boldsymbol{S}_h^t \equiv \sum_i S_{h,i}^{t_{k+1}} \varphi_i(\boldsymbol{x}).$$

With this reconstruction, Equation (5.32) defines an implicit method, where the solution at the new time step $S_{h,i}^{t_{k+1}}$ is defined implicitly by the solution of Equation (5.32). Implicit schemes are usually much more stable than explicit schemes, in particular for parabolic problems. However, solving Equation (5.32) may pose challenges in itself, in particular if long time steps are attempted.

A different approach to solving the saturation equation is to design grids that flow in time with the solution. One of these approaches leads to the Euler–Lagrangian schemes, where the test functions in the variational problem themselves solve a separate problem to ensure that the need for saturation reconstructions is minimized. A related approach is defined on the original form of the saturation equation, where one may also consider solving the saturation evolution on characteristics of the system, that is, the paths in space-time where the solution is constant. These methods are known as methods of characteristics. Both of these approaches have found limited application in practice due to the challenges in defining the correct test functions and characteristic paths in complex multidimensional domains.

A more practical compromise is found in streamline methods, which assume that the direction of the total flux vector changes slowly in time, and therefore those directions can be used to propagate the solution. These methods have enjoyed a wider success in practical implementations.

5.3 MULTISCALE NUMERICAL METHODS AND PARAMETER UPSCALING

So far we have emphasized the interaction of scales, both in the derivation of coarse-scale model equations and in the derivation of numerical methods. It is important to recognize that if we use a numerical method to discretize our coarse-scale equations, we will obtain a significantly different approximation than if we were to use the same computational grid to discretize the fine-scale equations. We also expect that our numerical approximation is more appropriate for the coarse-scale than the fine-scale equations, because the disparity between the grid resolution and the scale of the equations necessarily is smaller, and it is therefore easier to achieve a grid that is sufficiently fine to allow the continuous solution to be adequately resolved.

Nevertheless, there may still remain a scale gap between our coarse-scale equations and the resolution of our computational grid. This was pointed out when we discussed the Godunov method for the saturation equation in Section 5.2.3, where we noted that the primary source of error was not in the approximation to the flux, but in the reconstruction of a continuous saturation from the discrete representation. A similar situation exists for the approximations we presented to the pressure equation in Section 5.2.2, where the smooth, polynomial basis functions frequently associated with finite element methods do not lead to reconstructions that are sufficiently accurate. In this section we will discuss methods that seek to overcome these problems. Our focus in this discussion will primarily be the pressure equation.

5.3.1 Multiscale Reconstructions for the Pressure Equation

The key ingredient in our earlier presentation of methods for the pressure equation was the reconstruction of a continuous function from the discrete representation. This allowed us to evaluate the variational integrals

$$A_{i,j} = \left(\boldsymbol{k}\nabla\vartheta_j, \nabla\varphi_i\right)$$

that form the coefficients of the discrete linear system. In our example, the reconstructions were chosen as the simplest possible. This is a common choice in porous media flow simulation, due to the low regularity of the solution and the limitations of computational resources.

However, it is often the case that the solution varies even within a single grid cell. Consider again the integral that defines the coefficients $A_{i,j}$. It may be the case that the permeability $\boldsymbol{k}$ has variations on a scale that is even finer than the grid cell. When this is the case, the standard basis functions may have a shape that is inconsistent with $\boldsymbol{k}$. Let us briefly consider a problem in one dimension, with no source term. We integrate our model elliptic equation, Equation (5.20), directly, and define the solution as

$$p(x) \sim \int^{x} k^{-1}dx.$$

We see that if the permeability has low values, the pressure will vary steeply, while it will be near constant when the permeability takes high values. Thus, the pressure needs not resemble a piecewise linear function. As such, it is tempting to define a basis function that also varies like k^{-1}, hoping that this will give a better reconstruction:

$$\vartheta_i^{Ms}(\partial\omega_{i,1}) = 0; \quad \vartheta_i^{Ms}(\partial\omega_{i,2}) = 1; \quad \frac{d}{dx}\left(k\frac{d\vartheta_i^{Ms}}{dx}\right) = 0 \text{ in } \omega_i.$$

We refer to reconstructions using basis functions that solve a problem within each grid cell as multiscale basis functions, and denote them by superscripts *Ms*.

With this definition, we see that if p is a function that solves our model elliptic Equation (5.20), then since

$$p_h = \mathcal{R}_p^{Ms} \mathcal{C}_P^{FE} p$$

solves the model equation within each grid cell, and is equal to p at the boundary of grid cells, also p_h solves Equation (5.20). But for given boundary conditions, Equation (5.20) has a unique solution, and therefore $p_h = p$. Moreover, we can show that since the reconstructions are exact for the true solution, then the variational problem will only be satisfied if the reconstructed solution is exact. In other words, the finite element method gives the exact solution with this reconstruction.

We refer to numerical methods using basis functions that satisfy a finer-scale problem, rather than simple polynomials, as multiscale numerical methods. Multiscale numerical methods come with a cost: in the one-dimensional example above, it is as expensive to evaluate the integrals to determine the basis functions as it is to simply integrate the problem to determine the solution directly.

In multiple dimensions, the exactness of the multiscale approach does not hold, since in more than one dimension it is not obvious how to correctly define boundary conditions for the basis functions. However, the multiscale method can now provide a computational advantage over the original continuous problem, since the problems within each grid block may be solved independently. Furthermore, it will frequently be sufficient to use approximate multiscale basis functions in order to achieve acceptable computational accuracy on the coarse scale.

Practical Aspects of Scalar Multiscale Basis Functions

We recall that both the finite element method and the control volume method used scalar basis functions. Their multiscale variants are obtained by substituting a scalar multiscale basis function, which in multiple dimensions can be defined as

$$\vartheta_{i,j}^{Ms}: \vartheta_{i,j}^{Ms}(y_{i,k}) = \delta_{k,j}; \quad \vartheta_{i,j}^{Ms}(x) = b_{i,j}(x) \text{ for } x \in \partial\omega_i; \quad \nabla\cdot\left(k\nabla\vartheta_{i,j}^{Ms}\right) = 0 \text{ in } \omega_i.$$

Here, the Kronecker delta $\delta_{k,j}$ is defined as equal to 1 if $k = j$, and zero otherwise. Thus, the first condition states that the multiscale basis function is similar to the regular basis functions in that it takes the value 1 at one vertex of the cell, 0 at the other vertexes. The second specifies that the multiscale basis function takes on prescribed values $b_{i,j}$ along cell boundaries. Finally, internal to the cell, the multiscale basis function satisfies the equation we wish to solve. This leads to two practical problems: How do we chose the functions b? And how do we solve the differential equation?

The answers are usually provided by some heuristic expression for b, and some new numerical discretization internal to the grid cell to approximate the differential equation. In practice, these methods tend to be sensitive to the choice of b, to the extent where the multiscale approximation may in some cases have worse approximation properties than the standard, single-scale method. Furthermore, multiscale methods are naturally more expensive than usual finite elements because a separate numerical problem has to be solved on every grid cell.

Once multiscale basis functions have been computed within each element, they are associated with vertex basis functions that resemble their regular finite element counterparts. Thus,

$$\vartheta_k^{Ms} = \sum_{i,j:y_{i,j}=y_k} \vartheta_{i,j}^{Ms}.$$

We can then use the multiscale basis functions directly in the finite element method as given by Equation (5.24), with the reconstruction operator

$$\mathcal{R}_p^{Ms} \boldsymbol{P}_h \equiv \sum_k P_{h,k} \vartheta_k^{Ms}.$$

The main advantage of scalar multiscale methods appears when many similar problems are to be solved, as is typically the case in time-dependent simulations of flow in porous media. Then, if one reuses the majority of multiscale basis functions between computations, the benefit of the increased accuracy in both the discrete approximation and the continuous reconstruction may justify the extra cost invested.

Practical Aspects of Vector Multiscale Basis Functions

We recall that for the mixed finite element method, we needed a vector basis function. Indeed, this basis function becomes of much greater interest than the scalar basis function, since it is the flow field that may have significant structure, while the pressure is usually a smoother function. We recall that the RT0 basis functions were defined such that they have constant divergence $\nabla \cdot \boldsymbol{\vartheta}_j \in \underline{w_h}$, they are curl free (that is $\nabla \times \boldsymbol{\vartheta}_j = 0$), and they have a nonzero (constant) flux only over edge j, $\boldsymbol{\upsilon}_j \cdot \mathbf{v}_j = 1$. The two first conditions imply that $\boldsymbol{\vartheta}_j$ can be derived from a potential, so that the vector basis function satisfies

$$\nabla \cdot \boldsymbol{\vartheta}_j = q_1 \quad \text{and} \quad \boldsymbol{\vartheta}_j = -\nabla q_2,$$

inside each grid cell, for some potential q_2, and a piecewise constant q_1. However, if we compare to the problem we want to solve, Equation (5.25), we see that our RT0 basis function essentially solve a different problem. Furthermore, it is intuitive how to define multiscale basis functions consistent with Equation (5.25): they must satisfy

$$\nabla \cdot \boldsymbol{\vartheta}_j^{Ms} = q_1 \quad \text{and} \quad \boldsymbol{\vartheta}_j^{Ms} = -\boldsymbol{k} \nabla q_2,$$

internal to each grid cell. In practice, it is natural to solve these equations over a domain that at least covers the two cells that share the edge j, and the constant flux over this edge is then enforced only in an integral sense.

The multiscale reconstruction operator for the velocity space is then defined in the same way as for the standard velocity spaces,

$$\mathcal{R}_{\boldsymbol{u}}^{Ms} \boldsymbol{U}_h \equiv \sum_i U_{h,i} \boldsymbol{\vartheta}_i^{Ms}.$$

This can be substituted into the mixed finite element Equation (5.27).

The advantage of vector (compared to scalar) basis function-based multiscale numerical methods lies in the weakened dependence on imposed boundary conditions. This also allows for more natural generalizations to complicated grids. However, as we discussed in Box 4.2, multiscale methods tend to inherit by design, and sometimes even emphasize, the limitations of their fine-scale counterparts. For numerical multiscale methods, this applies to properties of the single-scale methods such as symmetry, positive definiteness, and monotonicity of the discrete linear systems.

5.3.2 Parameter Upscaling for the Pressure Equation

The multiscale approach is a rather recent take on the older problem of unresolved parameters. Going back to our one-dimensional example in Section 5.3.1, we might be inclined to modify our interpretation of permeability rather than our reconstruction, for example, define approximate coefficients

$$A_{i,j} = \left(\boldsymbol{K} \nabla \vartheta_j, \nabla \varphi_i \right),$$

where the $\boldsymbol{K}$ represents some upscaled permeability, which is constant on each cell. Indeed, from our one-dimensional example, we can easily be convinced that choosing the harmonic average

$$K_i = \frac{1}{|\omega_i|} \int_{\omega_i} k^{-1} dx$$

might be a good idea. Since the upscaled permeability is constant on each cell, evaluating the coefficients $A_{i,j}$ is also simplified. As was the case with multiscale basis functions, this approach is exact in one dimension. However, we again face challenges when we extend this concept to multiple dimensions; the harmonic average is not correct, since contrary to a single dimension, fluids can simply flow around an impermeable patch. Thus, our upscaled permeability must account for the geometric structure of the permeability variations, and we may need to solve a local flow problem to capture the correct upscaled permeability.

When a local flow problem is solved to upscale permeability, permeability upscaling becomes in principle equivalent to using a multiscale method to obtain a discrete system. It is also fraught with the same problems in terms of boundary condition dependence. However, upscaling the permeability has the distinct disadvantage that it does not provide a natural reconstruction operator. This becomes an issue if some information (such as the velocity field) is desired at the fine scale. The lack of reconstruction operator also makes the coupling to multiphase flow more complex, as upscaled mobility functions appear to be inherently strongly boundary- and history-dependent.

5.3.3 Other Approaches

So far, we have considered both numerical methods themselves, and their multiscale counterparts, in the language of the heterogeneous multiscale methods (HMM) introduced in Chapter 4. This is partly a matter of choice, as certainly we could have presented many of the same ideas in a more traditional form. However, we hope that by staying consistent to a multiscale way of thinking throughout Chapters 4 and 5, the reader recognizes the similarity of thought between the development of coarse-scale model equations, and the numerical approaches that can address related problems. We will use the remainder of this section to touch upon some related topics that are meant to complement the ideas presented so far.

If we consider the observation that multiscale numerical methods need a finer grid to compute the multiscale basis functions, we can ask if this multiscale framework is the most efficient way to approximate the equations on that finest grid. With this perspective in mind, a clear relation between domain decomposition preconditioners for the finest grid and the multiscale methods presented here can be found. The most useful implication is that it allows us to think of our multiscale approximation also as preconditioners, and the domain decomposition framework gives us the tools to start thinking about improving the design of reconstruction operators, either locally or globally, through iterative methods.

Another issue that comes to the front is adaptivity. Adaptivity in terms of choice of coarse models has been an implicit theme of Chapter 4, but adaptivity can also be thought of in terms of grid structure, choice of numerical approximation, effort spent in obtaining resolved basis functions, and reuses of existing basis functions for time-dependent problems. Formulating an integrated and robust approach combining all of these issues is still an open challenge.

To conclude this section, we return to the HMM, whose language we have borrowed liberally from the start of Chapter 4. The two most prevalent ideas of the original exposition of HMM are: (1) models of the same system at different scales may have different mathematical character, and (2) fine-scale models should, for many applications, only be solved in a small fraction of the domain. We have generously applied HMM as a framework to describe how vertically integrated models can be developed, which is a case where the fine-scale model is very different from the coarse-scale model. However, we approached this from a continuous perspective, and thus the solution to our fine-scale model is really given on the whole domain. Only once, when deriving the Godunov method, have we visited the idea of solving the fine-scale model within localized regions of the domain. Here, the reconstruction gave us the initial conditions for computing a flux over an edge, a computation that can be localized to exactly around the edge.

Clearly, these concepts and methods are general and can be applied in many different ways. We have provided herein only a superficial and introductory exposition of some of the basic ideas. However, the results we have developed can be applied to practical problems in CO_2 storage. We demonstrate three such applications in the next section.

5.4 APPLICATIONS TO REALISTIC DATA

To give a flavor of real numbers associated with CO_2 storage, and also to illustrate how some of the analytical and numerical solutions in the previous sections can be applied, we will here consider three realistic data sets. The first illustrates an application of analytical solutions during injection, for which we consider a suite of properties taken from formations typical of North America. The second applies numerical methods to a one-dimensional problem over the mega temporal scale, for which dissolution and capillary trapping of CO_2 are important. The example is representative of a CO_2 plume migrating under a sloping caprock, with idealized parameters based on North Sea data. The third example uses a numerical method to simulate using both realistic parameters and geometry over long time scales based on a heterogeneous data set for the Johansen formation under the North Sea. This formation is moderately heterogeneous, with an interesting topography, and numerical methods combined with a vertically integrated formulation allow us to resolve these features. We also explore the effect of convective mixing in such a large system.

5.4.1 Injection into Homogeneous Confined Formations

In this section, we use the analytical solutions for the injection phase developed in Section 5.1.1. In that section, we gave examples of how modeling choices, such as reconstruction using a capillary fringe or sharp interface representation, affected the predictions of our models. Here, we will consider different potential storage formations, and consider the impact of their different geological characteristics on the evolution of the CO_2 plume.

We identify the four end-member cases discussed in Chapter 3, representing injection into "deep" (3000 meters) and "shallow" (1000 meters) formations, in "cold" and "warm" basins. For typical salinity ranges, we obtain the ranges of fluid densities and viscosities found in Tables 3.1 and 3.2. We will use the mean values of the ranges from those tables in this section.

In addition to the fluid properties, we also need data for formation properties. Data for the relative permeability functions for CO_2–brine systems in formations underlying Alberta were also discussed in Chapter 3. Because we will use a sharp interface model in this section, we only consider the end point relative permeability values for CO_2, as well as the residual saturations of brine, both of which are summarized from the work of Bennion and Bachu in Table 5.1. Also included are typical permeability and porosity values for these types of porous media.

The remaining parameters of the model are chosen as follows. We consider injection into a formation with a reasonable volume, setting the thickness of all

Table 5.1 Formation Properties for the Formations Considered

	Sandstone shallow (Viking 1240 m)	Sandstone deep (Basal 2732 m)	Carbonate shallow (Wabamun 1357 m)	Carbonate deep (Nisku 2049 m)
Residual saturation brine	0.55	0.3	0.6	0.35
Endpoint rel. perm. CO_2	0.35	0.55	0.55	0.2
Permeability (mD)	10	10	50	50
Porosity (%)	10	10	5	5

Table 5.2 Plume Extent in Dimensionless Coordinates and in Meters for the Eight Cases Considered

	Sandstones				Carbonates			
	Shallow		Deep		Shallow		Deep	
	Cold	Warm	Cold	Warm	Cold	Warm	Cold	Warm
Plume extent (χ_0 equals λ)	7.2	10.5	4.6	3.5	11.3	16.4	1.7	1.3
Plume extent (r_0) (m)	4632	9132	2925	3174	8709	17172	2589	2809
Gravity number Γ	0.012	0.016	0.026	0.051	0.059	0.079	0.126	0.259

considered formations to 50 m. Further, we are motivated by injection from a single small- to medium-sized power plant, from which we expect that 1 million tones per year (MT/y) of emissions are captured. We choose a hypothetical review period for our single injection well of 15 years, considering this a typical time frame for good assessment of system performance.

For these data, the value of the dimensionless group Γ defined in Section 5.1.1 ranges between 0.012 and 0.26. We can therefore apply the simplified solution given by Equation (5.8) for injection of pure CO_2. In particular, we recall that for this simple model, the outer extent of the plume is given by the mobility ratio $\chi_0 = \lambda$. We use this relation to calculate the results in Table 5.2.

An interesting aspect of Table 5.2 is the enormous variability in the predicted plume extent. Intuitively, one might attribute this to the dependence of aquifer volume on porosity; however, this would be a mistaken conclusion, since the minimum plume extent is found in the carbonates, where the porosity is 50% lower than in the sandstones. Indeed, we see that the main effect comes from the mobility ratio, which critically affects how far the gravity tongue at the top of the formation can run ahead of the bulk CO_2. This is particularly true for the deep carbonates, where the plume extent is close to $\chi_0 = 1$, which is a piston displacement. This is strongly affected by the low value of end-point relative permeability for the injected CO_2, thereby highlighting an important effect of residual brine saturation.

Before we conclude this section, we again stress that these results are obtained using a simple immiscible sharp interface model. If miscibility was included, the extents would change by up to about 7%. The impact of using a fully developed capillary fringe model is of course dependent on the strength of capillary pressure; however, we expect from inspecting the figures in Section 5.1.1 that the impact may be on the order of 10–50% on the prediction of outer plume extent.

5.4.2 Long-Term Migration in a Sloping Aquifer

Since most real formations have some nonzero slope, we now consider the migration of CO_2 in mildly sloping formations. Here, we will use the Buckley–Leverett theory from Section 3.6 as well as more general numerical solutions to understand the CO_2 migration pattern. We will see the impact our model choices have on the resulting predictions. In particular, we will see how our predictions of the velocity of the plume tip depend on our model, and the duality between the simple expressions we can obtain from Buckley–Leverett theory and the results obtained by numerical simulations using more comprehensive models.

For a macroscale model based on incompressible, immiscible reconstructions for saturation with the Dupuit reconstruction for pressure, the saturation equation in one dimension for a homogeneous formation is derived following Section 3.6 and Section 5.1 as:

$$\Phi\frac{\partial S_c}{\partial t}+\frac{\partial}{\partial x_1}\left(\frac{\Lambda_c(S_c)}{\Lambda_\Sigma(S_c)}U_\Sigma-K\frac{\Lambda_\Pi(S_c)}{\Lambda_\Sigma(S_c)}\left[\frac{\partial P^{cap}(S_c)}{\partial x_1}+\Delta_\alpha\rho(\boldsymbol{e}_1\cdot\boldsymbol{g})\right]\right)=0. \quad (5.33)$$

Here, the first term among the three terms in the spatial derivative is the component due to the total flow of fluids (usually associated with the natural cycling of fluids in the subsurface), while the second and third terms represent the dispersive and advective components of the gravitational force acting on the CO_2 plume. In this section, we will for simplicity neglect the total flow, and assume that $U_\Sigma = 0$.

We start off with a brief analytical consideration of the expected velocity of the plume tip. Recall from Section 3.6 that we can approximate the evolution of the plume from a step-function initial condition based on the fractional flow function. This approach neglects the second-order term (derivative of coarse capillary pressure) in the saturation Equation (5.33). For sharp interface models, the coarse fractional flow function is strictly convex, so we do not have to worry about the formation of a shock. The tip speed is therefore simply the derivative of the fractional flow function evaluated at the coarse-scale tip saturation, $S_c = 0$:

$$\frac{v_{tip}\Phi}{K\Delta_\alpha\rho(\boldsymbol{e}_1\cdot\boldsymbol{g})}=\frac{d}{dS_c}\left(\frac{\Lambda_\Pi(S_c)}{\Lambda_\Sigma(S_c)}\right)\bigg|_{S_c=0}=\frac{d\Lambda_c}{dS_c}\bigg|_{S_c=0}=\frac{\lambda_c(s_{c,T})}{s_{c,T}}$$

or,

$$v_{tip}=\frac{K\Delta_\alpha\rho(\boldsymbol{e}_1\cdot\boldsymbol{g})\lambda_c(s_{c,T})}{\Phi s_{c,T}}.$$

BOX 5.1 *Analytical Approximations to Flow in a Sloping Aquifer*

In chapter 3 we reviewed constructing the Buckley–Leverett solution for one-dimensional problems where the second-order, capillary pressure, terms were negligible. One may argue that the problem of migration under a sloping aquifer can reasonably be approximating by the first-order terms, and therefore that the Buckley–Leverett theory is applicable. This is particularly appealing when using the sharp interface reconstruction, since the simple form of the coarse mobility functions allow for explicit expression of the solution to the initial value problem. However, for more complex reconstructions, closed form solutions become increasingly tedious to calculate. We have therefore chosen to use the problem of sloping aquifers to highlight the various implications of different reconstructions, while applying the numerical methods presented in this chapter.

We recognize that this expression simply expresses Darcy flow of CO_2 driven by a buoyant force $\Delta_\alpha \rho(\boldsymbol{e}_1 \cdot \boldsymbol{g})$. This indicates that the properties of the brine phase are of secondary importance for the tip speed (except for the density in the $\Delta_\alpha \rho$ term).

This simple tip speed calculation could of course also be applied for a macroscale model that includes capillary forces at the fine scale. However, then the coarse fractional flow function would no longer be convex, and we would need to consider the convex hull. With realistic fine-scale relative permeability and capillary pressure functions, this calculation quickly becomes intractable analytically. Further, we may be interested in the evolution not only of the tip, but also the plume itself. Therefore, we choose to use numerical solutions to study impact of these additional physical processes (see Box 5.1).

In this section, we have the advantage of only solving a simple equation, the saturation equation, since the pressure equation is trivial in one dimension (volume is conserved, therefore, total flow is constant). The saturation Equation (5.33) is written in the same form as that for which numerical solution approaches were developed in Section 5.2.3. We solve Equation (5.33) with the first-order explicit numerical method given in Section 5.2.3. The critical element of this method is to define a numerical approximation to the flux. We will do this by considering the two flux terms in Equation (5.33) individually.

The advective term (which is physically related to the upslope migration of CO_2) is evaluated by taking the saturation of each phase from the cell from which the phase is flowing. This is known as upstream weighting. Usually, this will be the downslope cell for the CO_2 and the upslope cell for the brine.

The most important aspect of the dispersive term (which is physically interpreted as the flattening of the plume) is the derivative of saturation. We approximate this as a simple first-order difference across the cell face. The saturations used in the mobility functions are still upstream weighted.

As an example of this approach, we chose data typical of a good aquifer. We define it to have a thickness of 50 m, a slope of 1%, porosity of 15%, and permeability of 10^{-13} m^2 (100 milliDarcy). The fluid properties are chosen based on the deep warm aquifer discussed in the previous example, giving densities of 1099 and 733 kg/m^3 for brine and CO_2, respectively, and viscosities of 0.5 and 0.06 mPa·s. We use the same fine-scale capillary pressure and relative permeability functions as in Section 5.1.1. We specify the residual brine and CO_2 saturations both as 0.2. For upscaled convective mixing rates, we use a value of 1.5 kg/m^2/y, which is chosen based on simulations reported by Riaz et al.

We are now prepared to calculate our numerical solution. As the initial condition, we chose to fill a length of 1 km up to the maximum saturation of CO_2. We chose a spatial resolution of 62 m, which due to stability requirements of the numerical method, limits our time step to about 3 months.

Figure 5.8 shows the numerical solution after 2000 years based on several different coarse-scale models. We have implemented three different saturation reconstructions (all use Dupuit pressure reconstruction): an immiscible incompressible reconstruction with a capillary transition zone (Section 4.5), simple sharp interface reconstrucion (Section 4.6), and a sharp-interface reconstrucion including convective mixing (Section 4.8). This last model will have a slightly different initial mass in the system than the other two models, due to dissolution. The upscaled convective mixing term, $\Psi_{\mathcal{M}^c_{\mathfrak{b}},\beta}\{\Xi\}$, is approximated as constant at 15 kg/m^2/y when $h_{C,B} > 0$.

We first calculate the hyperbolic expression for the tip speed which leads to an estimate of 6.3 m/y. This is in reasonable agreement with the numerical sharp interface solution, which predicts a tip speed of about 8 m/y. However, we see that this

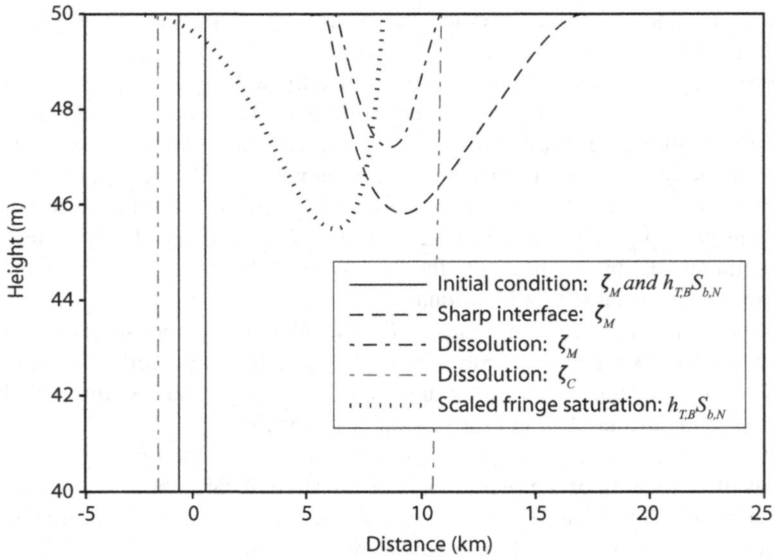

Figure 5.8 Simulation results after 2000 years for migration under a sloping caprock. Note that only the upper fifth of the aquifer is shown.

value is highly dependent on the choice of reconstruction in our coarse model. For our parameters, inclusion of either the capillary fringe or dissolution reduces the tip speed by a significant fraction.

It is equally interesting to consider the amount of CO_2 left in a mobile state, since this quantity correlates with risk. For the simplest model, with a sharp interface reconstruction and only residual CO_2 as a trapping mechanism, 49% of the CO_2 is still in a mobile state after 2000 years. Reconstructing a capillary fringe leads to more immobilized CO_2, thus the predicted amount of mobile CO_2 after 2000 years with a capillary fringe model is 44%. However, dissolution with convective mixing turns out to have a much greater impact, where after 2000 years only 15% of the original CO_2 is still in a mobile phase.

From this section we see how a simple numerical solution gives us quantitative results from which we can build an intuition about the relative importance of different physical processes included in the various reconstructions, as well as the calculation of important quantitative values such as plume tip migration speed.

5.4.3 Injection into the Johansen Formation

The Johansen formation is one of the potential target formations for CO_2 injection identified by the Norwegian government. It is located off the west coast of Norway, and is part of the deeper sedimentary structure that forms the Sognefjord delta. The formation itself has a horizontal extent of 100 km in the North-South direction, and 60 km in the East-West direction. Over this extent, the top surface varies between 2200 and 3100 meters depth, with formation thickness varying between 30 and 225 m.

We simulate a portion of the Johansen formation covering a horizontal extent of 2100 km^2. In Figure 5.9 we show the top surface of the simulation domain, together with a map of the thickness. The formation properties are taken from data provided by the MatMoRA project. We chose to use a sharp interface model, emphasizing the effect of including diffusion and convective mixing over the effects of the capillary fringe. A typical result from such a simulation is seen in Figure 5.10. In this figure, where CO_2 has been injected for 50 years followed by 200 years shut-in, we see the extent of free and residual phase CO_2, along with the impact of brine with dissolved CO_2. Note in particular the disconnected blobs of free phase CO_2. This is typical for such highly permeable formations, and is indicative of the importance of the geometry of the top boundary of the aquifer, where local dome structures serve to contain pockets of free-phase CO_2.

We are now also in a position to make a more quantitative version of the figure in the Intergovernmental Panel on Climate Change (IPCC) report, given as Figure 3.17 in this book. By letting our sharp interface, convective mixing model run past the injection period, we can look at the evolution of the CO_2 inventory, in terms of trapping mechanisms. We note some important aspects of Figure 5.11. First, the critical period for leakage is during the injection phase, and the following few centuries, when a significant portion of CO_2 is in free phase. After the first half

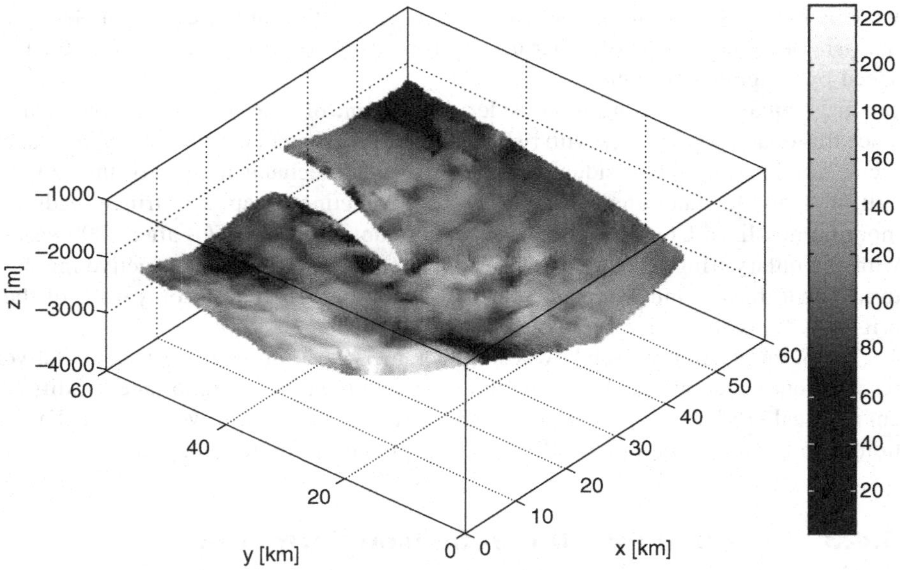

Figure 5.9 An elevation map of the computational domain taken from the Johansen formation. The gray scale indicates thickness of formation $h_{T,B}$ (figure adapted from Gasda et al. 2011).

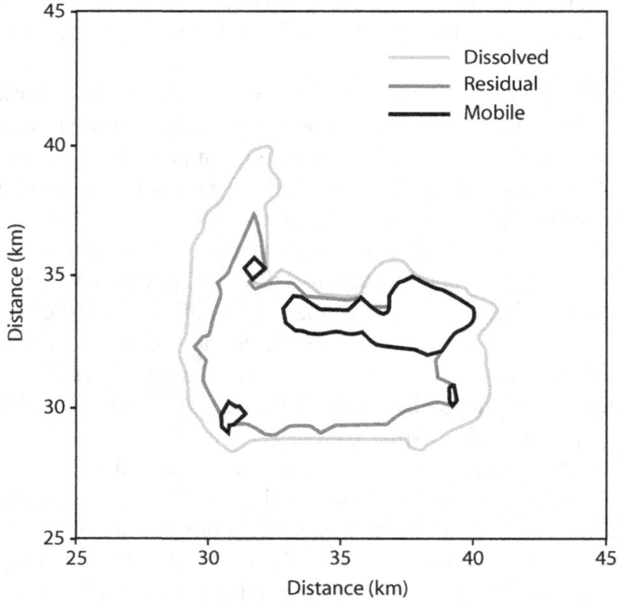

Figure 5.10 Distribution of free, residual, and dissolved CO_2 near the injection point. Note that this figure shows a subsection of the full simulation domain (from Sarah Gasda).

millennia, more than 95% of the CO_2 is either dissolved or residually trapped, and the risk associated with leakage is significantly reduced. Second, we note that virtually nothing happens after 1500 years. This is a consequence of the model and boundary conditions, since at late time, all free phase CO_2 is located under domes (structural traps) in the cap rock, with a completely CO_2 saturated brine below it, thus not permitting further buoyant flow or dissolution. With a more detailed reservoir characterization, we might be justified in specifying boundary conditions to the model that give a net flow of brine through the aquifer. This would transport away the dissolved CO_2, such that we would see a continuous dissolution process in the system, albeit at a relatively slow rate (see also Box 5.2).

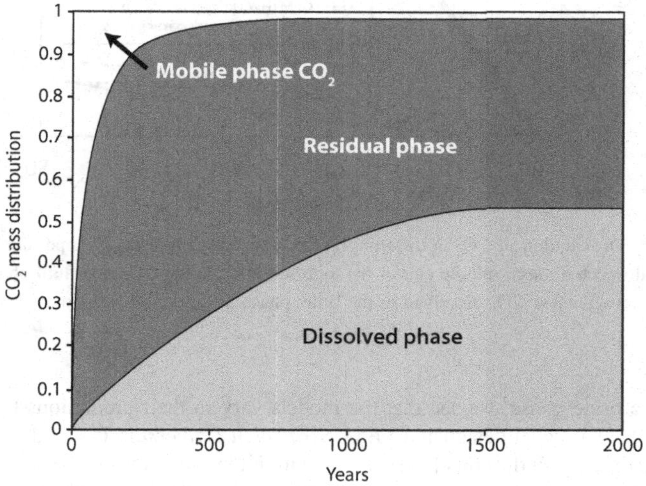

Figure 5.11 Fractional distribution of the CO_2 inventory. The intervals are sorted according to increasing risk of leakage, as suggested by the IPCC report.

BOX 5.2 *Comparison of Flow Predictions for the Johansen Formation*

The Johansen formation described in this section was also chosen as the basis for the last of three benchmark problems for a code intercomparison project led by the University of Stuttgart, Germany. For this study, the organizers provided problems with varying degree of physical and spatial complexity.

This particular benchmark problem had parameters and geometry taken from the central 8 by 8 km section of the full model. In this region, the participating groups were asked to model the fluid migration as a result of 15 kg/s CO_2 injection near the middle of the domain over 25 years, with a further 25 year postinjection period. As feedback, the participating groups submitted both integrated curves, as well as plots of saturation distribution.

When reviewing the integrated curves presented in Figure 5.A, it is natural to think that we have a pretty good understanding of how to model this problem, as most participants got very similar results. However, looking in particular at the mass of CO_2 dis-

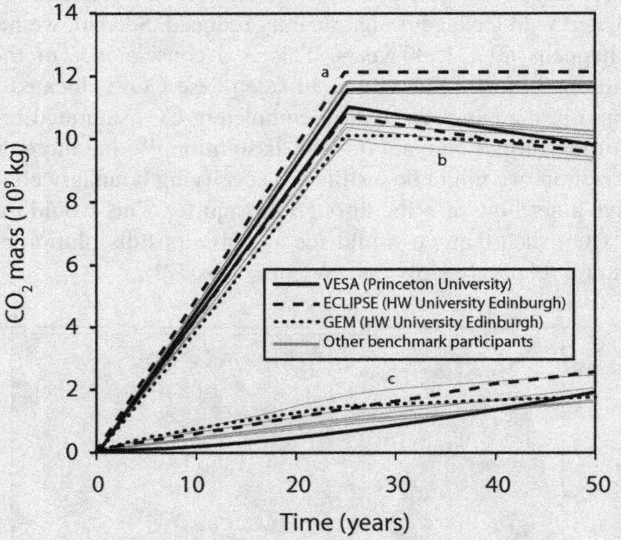

Figure 5.A Distribution of CO_2 in different phases as a function of time. Upper curves (a) represent total injected mass; middle curves (b) represent CO_2 in the CO_2-rich fluid phase, while lower curves (c) represent CO_2 dissolved in the brine phase. Figure modified from Class et al. (2009).

solved in the brine phase, we see that the models vary in their predictions by nearly a factor of two. The model designated *VESA (Princeton University)* is based on the macroscale models described in this book, and is within the range of predictions by the full 3D codes. Keep in mind, however, that the true solution to this problem may well be outside the envelope of the submitted approximate solutions, since most of the numerical methods introduce the same approximation (a coarse grid representation), which likely leads to a bias in the calculated results.

The disparity in the predictions of integrated measures such as mass distribution between phases can be understood more clearly when going into the detailed simulation results. This reveals that the actual numerical simulations predicted very different fluid migration patterns. At one end of the spectrum the University of Heriot-Watt University in Edinburgh applied the commercial simulator ECLIPSE in a simulation indicating that the CO_2 would migrate out of the domain. On the other hand, the same group using another commercial simulator (GEM), predicts that the CO_2 plume will migrate no more than a kilometer from the injection well during the simulation period.

While all the three-dimensional numerical methods applied are theoretically convergent to the true solution for this problem, this example shows that for practical problems, within reasonable constraints, it remains a challenge to calculate a good approximation. This again emphasizes the motivation behind the coarse-scale models developed in Chapter 4, where approximations are introduced explicitly using arguments of temporal and spatial scales, instead of implicitly through applying too coarse discrete representations of the fine-scale equations.

5.5 FURTHER READING

Books

CHEN, Z., G. HUAN, and Y. MA. 2006. *Computational Methods for Multiphase Flows in Porous Media.* Philadelphia: SIAM Press.

EFENDIEV, Y. and T.Y. HOU. 2009. *Multiscale Finite Element Methods: Theory and Applications.* New York: Springer.

LEVEQUE, R. 1992. Numerical methods for conservation laws. Lecture in Mathematics.

Papers

BLUNT, M. and P. KING. 1991. Relative permeabilities from two- and three-phase pore-scale network modeling. *Transport in Porous Media* 6, 407–433.

CLASS, H., A. EBIGBO, R. HELMIG, H. DAHLE, J.M. NORDBOTTEN, M.A. CELIA, P. AUDIGANE, M. DARCIS, J. ENNIS-KING, Y. FAN, B. FLEMISCH, S.E. GASDA, S. KRUG, D. LABREGERE, J. MIN, A. SBAI, S.G. THOMAS, and L. TRENTY. 2009. A benchmark study on problems related to CO_2 storage in geologic formations. *Special issue of Computational Geosciences*, 13(4), 409–434.

EIGESTAD, G.T., H.-K. DAHLE, B. HELLEVANG, F. RIIS, W.T. JOHANSEN, and E. ØIAN. 2009. Geological modeling and simulation of CO_2 injection in the Johansen formation. *Computational Geosciences*, 13, 435–450.

GASDA, S.E., J.M. NORDBOTTEN, and M.-A. CELIA. 2011. Vertically averaged approaches to CO_2 injection with solubility trapping. *Water Resources Research*, 47, W05528, doi: 10.1029/2010WR009075.

NORDBOTTEN, J.-M. and M.-A. CELIA. 2006a. An improved analytical solution for interface upconing around a well. *Water Resources Research*, 42, W08433.

NORDBOTTEN, J.-M. and M.-A. CELIA. 2006b. Similarity solutions for fluid injection into confined aquifers. *Journal of Fluid Mechanics*, 561, 307–327.

RIAZ, A., M. HESSE, H.-A. TCHELEPI, and F.-M. ORR. 2006. Onset of convection in a gravitationally unstable diffusive boundary layer in porous media. *Journal of Fluid Mechanics*, 548, 87–111.

Data

JOHANSEN FORMATION (MATMORA PROJECT). http://www.sintef.no/Projectweb/MatMorA/Downloads/

Chapter 6

Models for CO_2 Storage and Leakage

The CO_2 plume resulting from injection of the captured CO_2 emissions from a medium-sized power plant may have an areal extent that increases by several square kilometers per year of injection. The radius of influence, defined by pressure perturbations in the subsurface, is likely to expand even faster. This implies a large areal extent for the plumes, and suggests that different kinds of imperfections in the caprock may be encountered. These imperfections may allow for fluid leakage out of the injection formation and, therefore, must be taken into account. Such imperfections may include natural faults or fractures as well as old oil and gas wells. The problem of possible leakage along wells is especially important in North America, where a century and a half of oil and gas exploration and production has left millions of old wells in the ground (See Figure 1.7 and the associated discussion in Chapter 1). Comprehensive models for CO_2 storage must include these potential leakage paths. The presence of these kinds of leakage pathways provides strong vertical connectivity across formations. As such, our modeling needs to be expanded beyond the injection formation, to also consider other geological formations above and below.

In this chapter, we will discuss models involving a vertical sequence of geological formations. Our approach will use the large-scale models at the macro spatial and temporal scale developed in Chapter 4 to describe fluid flow within each permeable aquifer. The low-permeability aquitards (or caprock formations) separating the aquifers will also be modeled, providing a coupling between the aquifers. This coupling affects the pressure profile in the system as well as the fluid migration patterns and pathways.

Early examples of a single-aquifer single-aquitard model, including leakage across the aquitard, were presented in Section 2.4, where we looked at single-phase flow to a well in a leaky aquifer. The models in this chapter will consider an arbitrary number of permeable formations, separated by impermeable or low-permeability aquitards, with multiple leakage pathways and injection wells. We will consider both numerical and analytical solution techniques for these models.

Geological Storage of CO_2: Modeling Approaches for Large-Scale Simulation, First Edition.
Jan M. Nordbotten and Michael A. Celia.

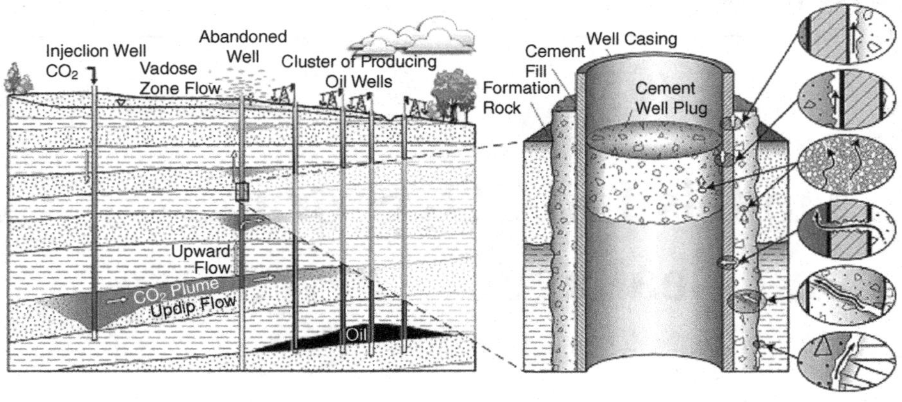

Figure 6.1 Schematic of the subsurface with CO_2 injection and potential leakage pathways due to the presence of old wells; the right panels shows potential nano- and microscale pathways for leakage along an abandoned well.

6.1 PROBLEM SETTING

We envision a general setting such as that shown in Figure 6.1. Our domain is on the spatial macroscale: we are concerned with the movement of both CO_2 and brine over tens of kilometers horizontally, and over a vertical succession of formations. In the figure, we have divided the subsurface into an alternating sequence of high-permeable aquifers and low-permeable aquitards. We assume that such a sequence can always be defined: if two aquifers (aquitards) were to be adjacent to one another, we could always merge them into a single aquifer (aquitard). Another feature of Figure 6.1 is the presence of multiple leakage paths. This restricts the applicability of purely analytical solutions such as those presented in Section 5.1. In their place, more involved approaches must be considered. Such approaches are developed in this chapter.

To provide a realistic context, we use specific data associated with formations in the Alberta Basin in western Canada. There, the sedimentary sequence overlying the bedrock involves on the order of 10–20 formations. Further, the density of existing oil and gas wells is on the order of ½ well per square kilometer, although it can be significantly higher in regions near oil and gas reservoirs. This implies that for a study of the region near an injection site, using, say, a 50 km by 50 km areal domain, we may expect to encounter more than a thousand wells. For a basin-wide study, to assess the joint impact of multiple injection sites, most or all of the more than 400,000 wells in the Alberta basin may be of concern. This kind of problem presents severe computational challenges, and requires the kinds of multi-scale models developed herein to provide practical solutions.

While our models will allow for inclusion of faults, our approach from a computational perspective emphasizes wells. Apart from their plentitude, wells are challenging in that: (1) the flow properties of an abandoned well are basically unknown or, at best, highly uncertain; (2) fluid flow along high-permeability wells may establish itself at very short time scales; and (3) the horizontal extent of wells is small, such that it is nearly impossible to spatially resolve all of them in a numerical simulation.

The combination of uncertain properties and high numbers (especially in North America) has led us to consider wells as one of the dominant leakage pathways as well as a dominant source of uncertainty. To address this uncertainty, we consider a stochastic framework where the dominant source of uncertainty corresponds to the well properties. Using an efficient implementation of the CO_2 leakage model as outlined in this chapter, we can consider a set of realizations of well properties, and look at the impact of the uncertainty range on the predicted leakage. This is demonstrated in the example calculations of Section 6.5.

6.2 COUPLED MACROSCALE MODELS

To develop appropriate multilayer models, we consider an aquifer within a general vertical sequence of alternating aquifers and aquitards. The aquifers are numbered with integers, starting with the bottommost aquifer. Let us consider an aquifer identified by the number l, and number the aquitard above it $l + 1/2$. In principle, there is no difference between aquifers and aquitards, as they are both porous media of similar geometric shape. However, flow in aquifers tends to be horizontal while flow in aquitards tends to be vertical (recall Box 2.4 on the tangent law), and this distinction needs to be kept in mind. While vertical flows might be seen as incompatible with vertical averaging, the reader should recall that the constraint we impose is that of a structured velocity in the vertical direction, defined at the mesoscale. In the following development, we will use the same ideas of compression as we used to model aquifers in Chapter 4 for both aquifers and aquitards. However, because aquitards have no confining formation above (the next layer in a succession would be an aquifer), have lower permeability leading to much longer characteristic time scales for flow, and the dominant flow direction tends to be vertical rather than horizontal, we are led to think differently about the choice of reconstruction operators for aquitards. We will discuss all of these issues in this section.

6.2.1 Coupling Aquifers and Aquitards

We recall the general form of our macroscale models for aquifers, given by the mass conservation Equation (4.6), together with suitably chosen reconstruction operators. For the moment, we will assume that we can treat the aquitards in the same manner, and our concern is therefore the coupling between aquifers and aquitards. The right-hand side of Equation (4.6) is given as a sum of sources and fluxes through the top and bottom surfaces, which takes the following form for aquifer number l and component i

$$\Psi^{i,l} = \int_{\zeta_B^l}^{\zeta_T^l} \psi^{i,l}\, dx_3 - \sum_{\alpha} \left[\rho_\alpha^l m_\alpha^{i,l} \boldsymbol{u}_\alpha^l \cdot (\boldsymbol{e}_3 - \nabla \zeta^l)\right]_{\zeta_B^l}^{\zeta_T^l}. \tag{6.1}$$

Here, we have added the superscript to denote aquifer number. In general, the source term $\psi^{i,l} = 0$, except at injection or producing wells, so that the first term on the right-hand side is zero almost everywhere. In contrast, the second term represents flow through the top and bottom boundaries. We denote these terms as in Chapter 3 by

$$\Psi^{i,l}_{\alpha,T} = \rho^l_\alpha m^{i,l}_\alpha \boldsymbol{u}^l_\alpha \cdot (\boldsymbol{e}_3 - \nabla\zeta^l)\big|_{\zeta^l_T} \quad \text{and} \quad \Psi^{i,l}_{\alpha,B} = \rho^l_\alpha m^{i,l}_\alpha \boldsymbol{u}^l_\alpha \cdot (\boldsymbol{e}_3 - \nabla\zeta^l)\big|_{\zeta^l_B}.$$

Thus, for each interface between an aquifer and an aquitard, we have new macroscale unknowns: the flux of each component in each phase through the interface corresponding to the top and bottom of each aquifer. Of course, these are the same fluxes at the top or bottom boundary of aquitards. We therefore need additional equations to couple aquifers to aquitards. At the fine scale, mass conservation across the interface between an aquifer and an aquitard implies

$$\sum_\alpha \Psi^{i,l}_{\alpha,T} = \sum_\alpha \Psi^{i,l+\frac{1}{2}}_{\alpha,B}.$$

Since $\zeta^l_T = \zeta^{l+1/2}_B$, we obtain the following constraint on our fine-scale fluxes:

$$(\boldsymbol{e}_3 - \nabla\zeta^l_T) \cdot \sum_\alpha \left(\rho^l_\alpha m^{i,l}_\alpha \boldsymbol{u}^l_\alpha - \rho^{l+\frac{1}{2}}_\alpha m^{i,l+\frac{1}{2}}_\alpha \boldsymbol{u}^{l+\frac{1}{2}}_\alpha \right)\Bigg|_{\zeta^l_T} = 0. \tag{6.2}$$

Additionally, the phase pressures must be continuous across the interface

$$p^l_\alpha(\zeta^l_T) = p^{l+1/2}_\alpha(\zeta^{l+1/2}_B). \tag{6.3}$$

The constraints given by Equations (6.2) and (6.3), together with equilibrium phase partitioning of the components, are sufficient to couple aquifers to aquitards. For the particular case of immiscible fluids, Equations (6.2) and (6.3) reduce to four equations for the four additional unknowns (two pressures, two phase fluxes normal to the interface).

6.2.2 Reconstruction Operators for Coupled Systems

The choice of reconstruction operators for coupled systems is complicated by the different time-scales of flow processes in aquifers and aquitards.

Equations (6.2) and (6.3) involve the pressure at the top of the formation as well as the normal fluxes of fluids. We recall from Section 4.7 that we needed a vertically structured reconstruction of pressure to obtain a non-zero reconstruction of vertical fluxes. Further, we emphasized in Section 5.1.2 that the Dupuit reconstruction is indeed only a special case of the vertically structured reconstruction, obtained when we considered the limit of large vertical permeability or negligible flow in the direction perpendicular to the formation boundaries. These ideas become of importance for the coupled systems.

Typical permeability contrasts between aquifers and aquitards are in the range of three orders of magnitude or more. Therefore, even though we need the concept of vertically structured reconstructions in both the aquifers and aquitards in order for Equation (6.2) to be helpful, in the aquifer the pressure will be close to fluid-static, and we can still approximate it using the Dupuit pressure reconstruction. Our interpretation of the Dupuit pressure reconstruction is therefore not that there is no vertical flow (vertical equilibrium), but only that the fluid pressures are close to fluid-static (that is, close to vertical equilibrium). This discussion implies that it is the vertical permeability of the aquitard, not the aquifer, which is important for fluid flow across the aquifer-aquitard interface. This conclusion is consistent with the discussion of flow perpendicular to layering in Box 2.1, where we noted that the harmonic average led to the correct effective permeability.

A second issue appears when considering saturation reconstruction. Since vertical flow in the aquitard is orders of magnitude slower than the horizontal flow in aquifers, and since there is no confining layer above an aquitard, it is not reasonable to expect that the lighter fluid accumulates at the top of the aquitard like it does for aquifers. Therefore, the saturation reconstructions of Chapter 4 is usually not appropriate for aquitards. Furthermore, as we have argued in Section 3.10, CO_2 often will not enter the aquitard because of capillary exclusion. For the case of capillary exclusion of CO_2 from the bulk aquitard rock, the CO_2 will only pass through the caprock in preferential flow paths that form leakage conduits, such as faults and wells. As such, the question of fine-scale saturation is primarily one of interest within leakage conduits.

6.2.3 Model Simplifications

So far in this section, we have argued that including multiple layers, consisting of aquifers and aquitards, is accomplished by treating all layers as detailed in Chapter 4, and enforcing the coupling through continuity of flux and pressure through the top and bottom boundaries. Within this general approach, we now discuss simplifications that are appropriate for our specific system. In particular, we will exploit the phenomenon of capillary entry pressure and the relatively small size of leakage conduits compared to our model domain.

In Section 3.10, we mentioned the phenomenon of capillary exclusion. This phenomenon ensures that for a wide range of conditions, CO_2 will not enter the aquitards. This range also correlates with the range of permissible operation of a CO_2 storage operation. We will, therefore, usually not need to consider models for reconstruction of the CO_2 saturation in the aquitard. For studies considering the consequence of breaching the capillary entry pressure into the confining aquitard, this assessment changes, and saturation reconstructions will need to be developed also for the bulk aquitards. In what follows, we will focus on leakage of CO_2 only through local pathways such as old wells. Brine will be allowed to leak through the bulk aquitard rock as well as the local leakage pathways.

Following the discussion of scales in Section 4.1, we recognize that leakage pathways are on the microscale and below, both in space and time. In particular, changes in saturation within a leakage pathway cannot be resolved on the coarse temporal scale. We will therefore not consider mass conservation strictly within leakage features, but rather exploit the scale separation and consider the saturation within a leakage pathway to be in equilibrium with respect to the coarse-scale variables. This requires a model for saturation in a leakage pathway, which we discuss below.

Finally, we point out that due to the difference in permeability between aquifers and aquitards, the flow direction of brine within the bulk aquitard rock will be largely perpendicular to the layering direction. We used this approximation already in the context of leaky aquitards in Chapter 2. It allows us to simplify significantly the models for aquitards. The important consequence of this approximation is that also $\nabla_{\|} \cdot \boldsymbol{U} = 0$ in aquitards; therefore, for incompressible flows, $\boldsymbol{U}$ is constant in the vertical direction. We may therefore think of flow in aquitards as essentially one-dimensional (in the vertical direction), with the one-dimensional velocity field being a function of the two horizontal space dimensions (i.e., the leakage through the aquitard can vary in space). This is consistent with most traditional models for single-phase flow with leakage.

6.2.4 Fluid Flow through Wells and Fractures

The dynamics of CO_2 and brine flowing up a well or fracture can be complex. In particular, if an open bore hole or fracture is encountered, it may be necessary to use more complicated models than Darcy's law, for example drift-flux models, to correctly capture the bubble flow of CO_2. Alternatively, if the temperature and pressure conditions are near the critical point for CO_2, significant thermal effects may be expected as the fluid migrates upward and expands. When these effects are expected to be important, a separate fine-scale model should be included for well-flow dynamics, using a multiscale framework such as presented in the previous chapter. In most cases, however, we expect fluid flow through even high-permeability wells and faults to be adequately modeled, within the large parameter uncertainties, by a linear, Darcy-like relationship. This is especially true when considering leakage pathways along the outside of the well casing—see Figure 6.1. We will, therefore, use Darcy's law to represent flow within or along wells. By using the assumption that wells are in steady state at the macroscale, such that the vertical flow is constant over the length of a well spanning an aquitard, we can average Darcy's law over the aquitard thickness to obtain

$$U_{\alpha,3}^{l+1/2} = -\Lambda_{\alpha,3}^{l+1/2}\left(S_{\alpha}^{l+1/2}\right)K_3^{l+1/2}\left(\frac{\widehat{p_\alpha}^{l+1}\left(\zeta_B^{l+1}\right)-\widehat{p_\alpha}^{l}\left(\zeta_T^{l}\right)}{h_{T,B}^{l+1/2}} - \varrho_\alpha^{l+1/2}\boldsymbol{g}\cdot\boldsymbol{e}_3\right). \tag{6.4}$$

Here, the upscaled coefficients are identified as

$$K_3^{l+1/2} \equiv \left(\int_{\zeta_B^{l+\frac{1}{2}}}^{\zeta_T^{l+\frac{1}{2}}} k_3^{-1}\,dx_3\right)^{-1} h_{T,B}^{l+1/2},$$

$$\Lambda_{\alpha,3}^{l+1/2} \equiv \left(\int_{\zeta_B^{l+1/2}}^{\zeta_T^{l+1/2}} \lambda_{\alpha,3}^{-1} k_3^{-1} \, dx_3 K_3^{l+1/2} \right)^{-1} h_{T,B}^{l+1/2}, \quad \text{and} \quad \varrho_\alpha^{l+1/2} \equiv \frac{1}{h_{T,B}^{l+1/2}} \int_{\zeta_B^{l+1/2}}^{\zeta_T^{l+1/2}} \rho_\alpha dx_3.$$

These expressions are of primary interest with regard to leakage pathways, but are also applicable to the aquitard itself.

6.2.5 Saturation Model for Leaking Wells

For a vertical segment of a leaking well, with length corresponding to the thickness of a formation, there is a balance between what flows into the well and what flows up the well. In a given aquifer, flow into (out of) the well from (to) the aquifer is balanced by the fluxes from below and above. Within the constraints of this mass balance, assignment of phase saturations within the leaky well is done in the context of local steady state. In particular, given the phase pressures at the bottom and top of a well segment, we need to determine the saturation in the well. Equation (6.4) provides two constraints on the balance equation, giving the bottom and top fluxes for a well segment along an aquifer. In addition, any of the upconing models presented in Section 5.1.2 present one further constraint, when we enforce that for two-phase flow to occur in the well, $h_{T,M}(r_w) = 0$, where r_w is the radius of the well. Below this threshold, CO_2 will maintain a finite thickness in the lower aquifer at the leaky well and therefore only CO_2 will flow in the well. For this case we have $Q_b = 0$, and $c_4 = 0$ in Equation (5.17). We can obtain an inequality for when brine starts flowing into the well by combining these two conditions in analogy to Equation (5.18). Thus, we will have no brine flow into the well, and hence the well is completely CO_2 saturated, if $J(0, r_{w,i}; c_4 = 0) = J(h_{T,M}, R_i; c_4 = 0)$, where J refers to the non-linear functions defining the solution in Equation (5.17). We write this out explicitly in terms of the sharp interface model as

$$R_i^2 < \left(r_{w,i}^2 - \frac{c_2}{c_1} \right) \exp\left(-\frac{c_1 h_{T,M}^2}{2c_3} \right) + \frac{c_2}{c_1}$$

When brine starts flowing with CO_2, we have four new unknowns compared to the single-aquifer case; the phase flow rates (which enter the coefficients c), the location of the outer boundary for the well model R_i, and the saturation in the well. Thus, we have five unknowns (fluid thickness in aquifer $h_{T,M}$, and the well model unknowns), and three constraints (Darcy's law for each phase, and the upconing model). Our approach is to use a model for the radius of influence of a leaking well to provide an equation for R_i (we will return to this in sections 6.3 and 6.4). The boundary condition for the upconing model is taken as a macroscale saturation in the aquifer, as determined using the boundary of the upconing model. This provides the fifth equation to close the system. The resulting system is nonlinear, but can be solved by iterative methods for nonlinear problems.

6.2.6 Saturation Model for Leaking Fractures

We do not know of any satisfactory saturation model for leaking fractures. However, a worst-case estimate in terms of CO_2 leakage can be obtained by assuming that $S_c^{l+1/2} = 1$ if $S_c^l > 0$. For leaky wells, our numerical studies indicate that a useful two-parameter model turns out to be the following,

$$S_c^{l+1/2} = \beta_1 (h_{T,M})^{\beta_2}$$

In numerical comparisons for leaking wells, the coefficient of proportionality as well as the exponent appear to be independent of time, but functions of the medium properties. This functional form may be a starting point for a coarse modeling of saturations in fractures.

6.2.7 Saturation Model for Continuing Wells

The saturation model for a leaky well can only be applied to formations where net flow enters the well in the formation directly below the lower aquitard associated with the aquifer of interest. This is often only the case for the lowest segment of the well. Further up the well, the saturation will be a function of the processes taking place inside the well, in addition to flow from the well into permeable formations. Currently, there is no good well model that models this behavior from first principles. However, the functional form

$$S_c^{l+1/2} = \left(S_c^{l-1/2}\right)^{\beta_3}$$

has appeared satisfactory in preliminary numerical experiments. The exponent $\beta_3 \geq 0$ will in general be a function of properties of the medium. A conservative estimate can be obtained by choosing the lower limit $\beta_3 = 0$.

6.2.8 Model Summary

The models we have presented in this section for mass conservation, fluid flow and saturation allow us to couple aquifers and aquitards. The modeling challenge is significantly greater for the coupled system than for the single aquifer. This is due to the importance of additional small scales (flow in leakage pathways) as well as the presence of larger scales (time scale of flow in aquitards). As a consequence, while modeling aquifers and aquitards is conceptually simple, we had to develop additional models to account for near-well flow into a well. This additional closure model is similar to the model for convective mixing in Section 4.9, in that it relies on physical understanding to obtain the general form, but requires fine-scale three-dimensional simulation to determine the actual parameters.

A final point is that since the flow in aquitards can be approximated as essentially one-dimensional, we can eliminate the aquitard by substituting the pressure reconstruction and using the fact that pressure is continuous over aquifer–aquitard interfaces. The resulting problem can therefore be written with only macroscale variables in aquifers as independent variables.

6.3 NUMERICAL SOLUTION

When multiple macroscale models are coupled together, the coupling represents both a virtual, and a real, third dimension. The real third dimension involves the different aquifers in the vertical succession, whose explicit inclusion is the aim of our modeling. The virtual third dimension appears in the structure of the equations for each of the aquifers. When considering the succession of aquifers and aquitards, we realize that this is simply a discretized vertical coordinate. However, just as we observed that our discretized macroscale equations are different from discretized microscale equations, the same holds for our system. The discretization appearing explicitly in our expressions for $\Psi_{\alpha}^{i,l}$ is the result of coarse-scale modeling, and the parameters entering the expressions reflect that. Further, the resolution of the discretization—one aquifer—is also the result of modeling, not an idea of a grid that resolves the micro-scale solution.

Therefore, while our domain is both virtually and in reality a three-dimensional domain, one of the dimensions has been discretized by modeling, thus our numerical approach is restricted to considering discretizations of the remaining two dimensions. Such models are sometimes referred to as "quasi-three-dimensional." Except for the wells and fractures, any standard approach will suffice. The simplest approach is to consider a splitting such as the IMPES method reviewed in Chapter 3. For the pressure and saturation equations, the methods of Section 5.2 can then be applied independently. More generally, fully implicit, or adaptively implicit strategies can be implemented if stability problems with the IMPES formulation are found to be limiting in computational performance.

The numerical grid should represent the features of the problem, and in our case a dominating feature is the presence of wells and fractures. Their presence leads to localized discontinuities in the aquitard vertical permeability $K_3^{l+1/2}$. We have two options in order to capture the effect of such discontinuities—either through local refinement in the two-dimensional grid or through specialized methods that handle these specific features. In the context of wells, we have chosen the second approach.

6.3.1 Well Module for Coarse Grids

In order to avoid excessive grid refinement near wells, we consider embedding a local representation of the well within a coarse grid block. To develop this local representation, we consider a single grid block, within which a well is located. The horizontal fluxes can be discretized using any of the numerical methods in Section 5.2. Further, we can establish the saturation in the well segment using the approach of Section 6.2, where we take the radius of the upconing solution to be the effective radius of the grid block,

$$R = \sqrt{\frac{|\omega_i|}{\pi}}$$

where $|\omega_i|$ denotes the area of grid cell number i. Our interest is in using the upconing model based on the vertically structured pressure reconstruction, since this most accurately captures the flow near the well. This implies a sharp interface approximation, and we approximate the interface elevation at the boundary by the average saturation in the grid block

$$h_{T,M}^R \approx h_{T,B,i} \frac{S_{c,h,i}}{\max S_{c,h,i} - \min S_{c,h,i}}.$$

Here $\max S_{c,h,i}$ and $\min S_{c,h,i}$ represent the maximum and minimum brine saturations in grid block i that can be achieved according to the coarse-scale capillary pressure functions. Thus, these limits are the coarse-scale residual saturations.

To complete our well module, we need a relationship between the grid-block pressure and the pressure at the well. This is achieved through a so-called Peaceman-type approach, named after a seminal paper published in the oil industry literature by Donald Peaceman from Exxon. In this approach, we use the grid-block size, the flow rate to the well, and the fluid distribution to calculate the difference between the average grid-block pressure and the pressure at the well. This is accomplished using the macro-scale Darcy's law in radial coordinates (see e.g., Eq. (5.9)) to calculate the variation of the pressure from the effective radius of the grid block to the well. This pressure profile must then be integrated again to obtain an expression involving average pressure, so that we can eliminate the unknown pressure at the effective radius. In practice, this approach leads to integrals that are too cumbersome for all but the Muskat upconing model.

In summary, a model for wells embedded in coarse grid blocks can be constructed by using the upconing model based on vertically structured pressure reconstruction to obtain saturation in wells, and integrating the Muskat upconing model to obtain a Peaceman-type correction to pressure.

6.4 SEMI-ANALYTICAL SOLUTIONS

Our macro-scale formulations have provided us with models that are much simpler than the underlying microscale models, in terms of both physical complexity and smoothness of solution. In Chapter 5, we considered both analytical and numerical approximations to the macro-scale models. In the previous section, we saw how these numerical approximations extend to the coupled system involving multiple porous formations, coupled through both low-permeability aquitards and more highly permeable features like leaky wells.

It is natural to ask whether any analytical solutions can be obtained for the coupled system. The answer is multifaceted, in that certain approximations that have an analytic character can be obtained. However, as in Chapter 5, those solutions require additional assumptions about the system to allow for analytical tractability.

The overall result for our multi-aquifer, multi-well system is a semi-analytical solution framework, in which analytical solutions in space are combined with discrete time-stepping to accommodate the nonlinearities of the system. We outline this semi-analytical approach in this section. We also comment that when the assumptions required for these solutions are not met, the semi-analytical approximations may still be applied; however, the expected accuracy of the approximation deteriorates. In many ways, this reflects practical modeling in most sciences: relationships are derived in controlled systems (lab experiments, analytical solutions, or numerical methods for sufficiently smooth problems), but in applications to real-world problems, the system is often outside the range of validity of at least some of the assumptions. In such cases, the analyst must then use best judgments to assess the impact and thereby evaluate the utility of the model.

In this section, we will emphasize semi-analytical solutions for pure brine systems as well as CO_2–brine systems. The importance of pure brine systems lies in the fact that even under the most extensive CO_2 storage scenarios, the fraction of CO_2 in the majority of formations is likely to be low, and the much simpler analysis involved with a single fluid is likely to provide useful estimates of important variables such as radius of influence of pressure propagation and large-scale alterations in groundwater flow.

6.4.1 General Assumptions

The blanket assumptions we will make throughout the section are: (1) aquifers are essentially horizontal, so that $\boldsymbol{e}_3 = \boldsymbol{e}_z$ and ζ_B^l, ζ_T^l, and ζ_P^l are parallel; (2) within each formation, we assume that fluid properties (viscosity, density and compressibility) are constant; (3) within each formation, we assume that material properties (porosity and permeability) are constant, homogeneous and isotropic in the horizontal directions; (4) we do not consider fractures; and (5) we do not consider inter-phase mass transfer (miscibility) in these models.

For convenience, we will define the pressure datum to correspond to the bottom of each aquifer, such that $\zeta_P^l = \zeta_B^l$. And for ease of notation we will omit the $\|$ subscript from the differential operator.

6.4.2 Impermeable Aquitards, Pure Brine

We first consider a system with only brine, with impermeable aquitards so that the only communication between aquifers is through wells. This system is completely determined by the evolution of a pressure equation, which takes the form

$$C^l \frac{\partial P^l}{\partial t} - \nabla \cdot \left(\Lambda^l K^l \nabla P^l \right) = \Upsilon^l \tag{6.5}$$

Here and in the rest of this subsection, we have omitted the subscript b since we only have the brine phase. Further, Equation (6.5) is simplified using the usual assumption leading to a linear equation (see also Section 2.2.2).

Considering the large disparity in scales between the macro scale (which is the scale at which we write our equations) and the microscale (which is the scale of the wells), we will model the wells as point sources and sinks in the macroscale model. However, we need to consider the finite radius of the well, when we consider the impact of leakage from a well on the pressure at its own location. This will become clear subsequently. Our right hand side then takes the form

$$\Upsilon^l = \delta(\boldsymbol{x}-\boldsymbol{x}_0)Q_0^l + \sum_{i=1}^{N_w} \delta(\boldsymbol{x}-\boldsymbol{x}_i)Q_i^l \tag{6.6}$$

where we have denoted the injection well by subscript 0, with injection rate given by Q_0^l, and each of the total of N_w potentially leaky wells has flow rates given by Q_i^l. Since the equations are linear, multiple injection wells can be accommodated by superposition. We have also used the Dirac delta distribution, using $\boldsymbol{x}_i$ to indicate the position of wells. The flow rates from the passive (or "abandoned") wells are not known a priori but can be obtained from Equation (6.4) as

$$Q_i^l = \frac{1}{\varrho^l}\Delta_3\left(\varrho^l \Lambda^l \bar{K}_i^l\left(\frac{\Delta_3 P^l}{h_{T,B}^l} + \left(\varrho^l - \varrho^{l-1/2}\frac{h_{T,B}^{l-1/2}}{h_{T,B}^l}\right)g\right)\right). \tag{6.7}$$

We have introduced the notation Δ_3, which is a difference operator with respect to the aquifer numbering, $\Delta_3 f^l = f^{l+1/2} - f^{l-1/2}$. The total well permeability $\bar{K}_i^{l+1/2}$ represents the horizontally integrated macroscale permeability of the well. Given a well radius $r_{w,i}$, we have the relation $\bar{K}_i^{l+1/2} \approx \pi\left(r_{w,i}^{l+1/2}\right)^2 K_z^{l+1/2}(\boldsymbol{x}_i)$. While the fluid properties are constant within each aquifer, we allow properties to vary from formation to formation.

We recall from Chapter 2 that for a constant-rate pumping well in a confined aquifer, the pressure response is given by

$$\Delta_{w,i}P^l = \frac{Q_i^l}{4\pi K^l \Lambda^l} W(\chi_i),$$

where we define $\Delta_{w,i}P^l$ as the pressure change due to injection at well i. The dimensionless group is defined as

$$\chi_i = \frac{r_i^2 C^l}{4\Lambda^l K^l t}$$

and the function $r_i(\boldsymbol{x}) = \max(|x - x_i|, r_{w,i})$ is used to indicate the distance from the center of the well, excluding positions inside the well. Now, for a time varying rate, such as will be the case for leaking wells, the linearity of the system implies that pressure response is given by a convolution integral,

$$\Delta_{w,i}P^l = \frac{\left(\frac{d}{dt}Q_i^l\right) * W(\chi_i)}{4\pi\Lambda^l K^l}. \tag{6.8}$$

The convolution operator is defined for functions a and b as

$$a * b \equiv \int_0^t a(t')b(t-t')dt'.$$

By superposition, the pressure at any point in the system is then given by

$$P^l(\boldsymbol{x},t) = P^l(\boldsymbol{x},0) + \sum_{i=0}^{N_w} \Delta_{w,i}P^l = P^l(\boldsymbol{x},0) + \sum_{i=0}^{N_w} \frac{\left(\frac{d}{dt}Q_i^l\right) * W(\chi_i)}{4\pi\Lambda^l K^l}. \tag{6.9}$$

We now proceed by considering the pressures at the abandoned wells in each aquifer, $P^l(\boldsymbol{x}_i)$, as our unknowns. If we have N_{Af} aquifers, and all wells are perforated in all aquifers, then this would lead to $N_w \cdot N_{Af}$ unknowns. By evaluating Equation (6.9) at these $N_w \cdot N_{Af}$ locations, we have as many equations as we have unknowns to propagate the system. The Darcy expressions along each well segment, given in Equation (6.7), form constitutive relations. Note that this reformulation of Equation (6.5) is almost exact, no approximations apart from the scale separation between the well and the domain have been introduced, since we have simply integrated the spatial differential operator analytically.

Equations (6.7) and (6.9) form a system of convolution equations, due to the presence of the convolution integral. This system can be solved by any appropriate time-stepping method. However, every time step will in general require more computation than the previous, since an integral over a successively longer time interval must be evaluated. Thus, the computational time increases more than proportionately to the simulated time, which can be a severe limit on practical computations. There are three approaches to simplify this computational hurdle.

Solutions in Laplace Transform Space

Since our equations are purely differential in time, we can consider application of the Laplace transform. This leads to a system of algebraic equations which can be solved in the transform space. However, the inverse Laplace transform cannot be evaluated analytically for this problem, and approximate inversion is difficult, even in the case of small numbers of leaking wells.

Volume Preserving Approximation of Convolution Integral

The convolution integral defining $\Delta_{w,i}P^l$ represents the pressure buildup due to the time history of flow to the well segment. If we consider a constant-rate injection well, then it is reasonable to assume that the leakage will increase monotonically in time. To simplify the convolution integral, we may therefore think of approximating the time history of flow to a well as constant at the current rate, dating back to a time $\tilde{t}_i^l$, constrained so that the correct total volume is preserved. This time is therefore defined as

$$\tilde{t}_i^l = t - \frac{V_i^l(t)}{Q_i^l(t)}$$

The volume V_i^l is defined as

$$V_i^l(t) = \int_0^t Q_i^l(t')\,dt'.$$

We now define the approximate time history at a time $t' < t$ by

$$\hat{Q}_i^l(t') = \mathcal{R}_Q\{Q_i^l(t), V_i^l(t)\} \equiv \begin{cases} Q_i^l(t) & \text{if} \quad t' \geq \tilde{t}_i^l \\ 0 & \text{if} \quad t' < \tilde{t}_i^l \end{cases}.$$

Using this approximation, the time derivative in the convolution integral becomes a Dirac distribution, leading to a simple expression for the integral. We obtain the following expression

$$\Delta_{w,i}P^l \approx \frac{\left(\frac{d}{dt}\hat{Q}_i^l\right) * W(\chi_i)}{4\pi\Lambda^l K^l} = \frac{Q_i^l}{4\pi\Lambda^l K^l} W\left(\frac{r_i^2 \Phi^l C^l}{4\Lambda^l K^l \left(t - \tilde{t}_i^l\right)}\right).$$

With this approximation, the evolution equations for the problem do not contain convolutions in time. However, the history of the problem must still be integrated to get an accurate estimate of $V_i^l(t)$. Thus, a time-stepping method is still needed; however, the computational complexity is now the same for all time steps, so that the computational time increases linearly with the simulation time.

This volume preserving approximation of the convolution integral is the approach we later extend to systems with both CO_2 and brine.

Self-Similar Approximation of Convolution Integral

We again consider the case of a constant injection rate. Then, as mentioned earlier, the flow to any leaking well can be expected to be close to a monotonic function, since the driving pressure gradient from the injection well will increase monotonically in time. Also, we observe that the well function is strictly increasing. Therefore, it is natural to consider the following as an upper bound on the convolution integral

$$\frac{\left(\frac{d}{dt}Q_i^l\right) * W(\chi_i)}{4\pi\Lambda^l K^l} \lesssim \frac{Q_i^l W(\chi_i)}{4\pi\Lambda^l K^l}.$$

Further, observations of solutions for simple system (using either the Laplace transform method or a time-stepping method for the full convolution integral to solve for a single leaky well) indicate that for intermediate times the solution scales in a way such that the historic time $\tilde{t}_i^l$, defined in the previous section, is approximately given by $\tilde{t}_i^l \approx 0.08t$. This allows us to approximate the pressure response without knowing the history of the problem,

$$\Delta_{w,i}P^l \approx \frac{Q_i^l}{4\pi K^l \Lambda^l} W\left(\frac{r_i^2 C^l}{0.92 \cdot 4\Lambda^l K^l t}\right).$$

This expression allows for the solution at any time to be approximated by a system of linear equations, without using any time-stepping.

While not reviewed herein, this approach can also be extended to CO_2–brine systems. However, the saturation equation forces a time-stepping procedure to be introduced, voiding the motivation and advantage of this approximation.

Approximations to the Well Function

In practice, the full well function is always evaluated by a series expansion. As mentioned in Section 2.5, retaining only two terms in this series is equivalent to a so-called Cooper–Jacob expansion. This is also the approximation with the clearest physical interpretation, as it is equivalent to solving the incompressible equations (whose solution is a logarithm) inside a domain that expands outward at a rate proportional to the square root of time (this defines the radius within which the Cooper–Jacob approximation is valid).

In general, using only two terms in the expansion gives reasonable results apart from early time. At early time, the approximation can be improved using any choice of an even number of terms in the expansion for the well function, bounded below by zero (expansions with an odd number of terms diverge to plus infinity without crossing zero, and are therefore more difficult to handle).

Comment on the Linear System

The system formed by Equations (6.7) and (6.9) is a linear system of equations for the unknowns $P^l(\boldsymbol{x}_i)$. Consider the equations and well segments as being numbered within aquifers first, then vertically. Then the system matrix will be block tri-diagonal with blocks of size $N_w \times N_w$. The main tri-diagonal set of blocks will fill up, since each aquifer connects to an aquifer above and below, and the remainder of the system matrix will be empty. The system matrix is not symmetric. When not all wells perforate all aquifers, the system will consist of blocks of size $N_w^l \times N_w^k$, where N_w^l denotes the number of wells perforating aquifer l. The general structure remains the same.

If a truncated series expansion is used for the well function, every well will have a time-dependent radius of influence. This will lead to additional sparsity in the system matrix for early times.

6.4.3 Permeable Aquitards, Pure Brine

The case of permeable aquitards is treated similarly to the impermeable aquitards, except that more complex interactions are introduced due to the diffuse leakage of brine across the aquitard formations. Again, the approach is to consider the system by superposition. We recall from Section 2.4 that the hydraulic head response for a single pumping well in a system with two aquifers separated by one permeable aquitard can be written in terms of Bessel functions. These solutions can be expanded for a more general system of multiple aquifers and multiple aquitards. While the resulting expressions are somewhat more involved, we will keep the notation as simple as possible through use of the following fundamental solution,

$$\Delta_{w,i,k}P^l = \left(\frac{d}{dt}Q_i^l\right) * W^{l,k}(r_i, t).$$

Note that we have expanded the delta notation, so that $\Delta_{w,i,k}P^l$ is the change in pressure in formation l due to flow into formation k at well segment i, and $W^{l,k}$ is the associated well function. This implies that the flow into a formation l at well segment i can affect the pressure in all other formations, not only in the host formation. Consequently, when applying superposition to find the pressure at a point we obtain, instead of Equation (6.9), the following equation:

$$P^l(\boldsymbol{x}, t) = P^l(\boldsymbol{x}, 0) + \sum_{k=1}^{N_{Af}} \sum_{i=0}^{N_w} \Delta_{w,i,k}P^l = P^l(\boldsymbol{x}, 0) + \sum_{k=1}^{N_{Af}} \sum_{i=0}^{N_w} \left(\frac{d}{dt}Q_i^l\right) * W^{l,k}(r_i, t).$$

Again we eliminate the well fluxes using the vertical Darcy equation, and write the system only in terms of the pressure at the well segments. Now the computational challenge again lies in simplifying the convolution integral, and any of the methods from the previous section can be attempted.

The series expansion for the multi-aquifer well function has a different character than the confined aquifer well function. Most notably, it arises from an eigenvalue expansion, and converges rather slowly. Alternatively, one may use the physical interpretation of the Cooper–Jacob approximation to design an approximation to the multi-aquifer. This involves consideration of a stationary solution inside a moving outer boundary.

The linear system arising for this model is no longer block tri-diagonal, instead it is a full matrix. For more details on the fundamental solution for systems with permeable aquitards, see Hunt (1985).

6.4.4 Impermeable Aquitards, CO_2, and Brine

We briefly describe the procedure for solving CO_2 and brine systems by considering it as an expansion of the pure brine model. The expansion involves inclusion of a saturation model and the adaptation of the well functions to account for the spatial and temporal variability of fluid saturations within the domain. We will also need to make some additional assumptions, as compared to the single-phase brine models. These will be presented and discussed in the context of the derivations that follow. We will only consider a single injection well in order to keep the notation simpler. In principle, multiple injection wells can be included in the model, although some additional complications associated with overlapping CO_2 plumes arise in the formulation.

Representation of Mass

For each well segment in an aquifer, we denote by $M_{\alpha,i}^l$ the mass of phase α associated with that well segment, where the mass will be in the form of a plume emanating from that well segment into the adjacent formation. There is no other representation of mass in the system, so mass conservation implies that

$$\sum_{i,l} M_{\alpha,i}^{l} = \int_0^t \varrho^l Q_{\alpha,0}^l \, dt' = \varrho^l V_{\alpha,0}^l(t),$$

where $Q_{\alpha,0}^l$ is the injection rate and $V_{\alpha,0}^l(t)$ is the cumulative volume. Mass associated with one well segment, which forms a plume around that segment, may leak and become associated with a different well segment. This will typically be the case for the injection plume, which will spread until it leaks into another permeable formation via flow along one or more of the leaky wells. Similarly, a plume associated with a segment of a leaky well may leak further up, either through its own well, or by any other well it may contact. However, mass can only be associated with an aquifer well segment if it has arrived there through that same well. This allows us to write the following mass balance expression,

$$\frac{\partial}{\partial t} M_{\alpha,i}^l = \varrho^{l-1/2} Q_{\alpha,i}^{l-1/2} - \sum_{j \in \mathcal{D}_{\alpha,i}^l} \varrho^{l+1/2} Q_{\alpha,j}^{l+1/2}. \tag{6.10}$$

Here we denote well segments draining from the plume of segment i in layer l as the set $\mathcal{D}_{\alpha,i}^l$. We will come back to this set later.

Reconstruction of Saturation in Aquifers

A continuous saturation field is reconstructed from the discrete masses $M_{\alpha,i}^l$. We assume that a plume can be reconstructed for each well segment, and that this plume will have local radial symmetry around its well segment. Then, for segments that have a positive mass of CO_2, we can use the self-similar representations of injection plumes given in Section 5.1. We apply the same approximation as for the pure brine system in Section 6.4.1, and approximate the time-dependent history of the well flow rates $Q_{\alpha,i}^l$ by constant rates over a time interval determined from the total volume that has entered the formation through that well, $V_{\alpha,i}^l$. The self-similar solution framework then gives the plume, and hence the fluid distribution, around the well segment.

Note that this approach leads to conceptual difficulties if $Q_\alpha < 0$, because the self-similar solution does not apply for this case. Depending on the magnitude of the negative fluid flow, the modeler must decide whether the self-similar shape will still provide an acceptable approximation, or whether other models should be applied. In practice, we have found that these cases occur rarely during the injection phase, and do not account for a significant contribution to the overall fluid migration. However, these aspects lead to challenges when considering this semi-analytical approach for the post injection dynamics.

Reconstruction of Saturation in Wells

The approach for reconstructing a saturation in a well segment through an aquitard follows the modeling approach presented in Section 6.2. However, for the semi-analytical approaches, a complication arises in that there is no clear length scale, such as the effective radius of the grid, from which to define the outer radius for the upconing solution.

The simplest approach is to consider that there is an outer boundary, and use the limiting relation given in Equation (5.18) in place of a more complex upconing model. This will likely be a sufficiently accurate approximation for most practical cases. The interface height boundary in Equation (5.18) is naturally interpreted as the saturation at the well in the absence of upconing.

To obtain improved accuracy, one may construct models for the radius of influence, so that the upconing solutions can be applied. For single-phase flow, the radius of influence may be estimated as

$$r_{infl} \sim \sqrt{k_r}\left(\frac{\Sigma_\alpha Q_\alpha}{|U| h_{T,B}^2}\right)^{\alpha_{infl}}.$$

Here U is the background flow rate (as a result of the influence of other wells as well as background regional flows), while the exponent $\alpha_{infl} = 1$ is an appropriate choice for single-phase flow. For two-phase flow, we obtain better results, in terms of comparison with resolved three-dimensional simulations on simple systems, using an exponent of 2.

Well Functions

Recall from the self-similar solutions for CO_2 injection that the global pressure was proportional to the injection pressure at the well. This proportionality is what we exploited as superposition in the single-phase models. While the compressible two-phase model equations are no longer linear, the pressure equation is linear for a fixed saturation distribution in space. Therefore, we will approximate the solution to the equation for global pressure by superposition, in the form of Equation (6.9). However, instead of the single-phase well functions, we will use a modified construction to account for the higher mobility of CO_2 near the well. Note that all our well functions are approximate, since the expressions are constructed to be locally radially symmetric and, therefore, they neglect non-symmetric features of the saturation distribution. We consider three cases for corrections to the well function to account for fluid saturations within the formation.

No CO_2 Near Well For this case, we apply the single-phase well functions.

Well Has Associated Mass of CO_2 For this case, we apply the single-phase well function outside the region of reconstructed CO_2 saturation. Inside this region, we integrate the similarity solution using Equation (5.9), with the pressure from the single-phase region as a boundary condition. Note that this approach neglects compressibility inside the CO_2 region, which is consistent with the Cooper–Jacob two-term approximation to the well function.

Well Is within the Range of the Plume from a Different Well For this case we construct a radially symmetric approximation to the saturation field. This is most simply achieved using an expression of the form $\tilde{S}(r_i) = \|S(r_i, \theta)\|_\theta$, in some suitable norm, where the idea is to consider the saturation around the full circle (i.e., in the θ-diretion, defined relative to the well of interest) and choose a suitable representation to assign a function that does not vary in θ. Our experience is that simply using the inf norm (i.e., taking the minimum value over θ) gives satisfactory results. Once an approximate saturation representation has been obtained, the well function can be obtained as in the previous case.

Draining Sets

For the mass conservation equations at well segments, Equation (6.10), we need to define the draining set for a well i, defined as the set containing the index of neighboring wells that drain CO_2 from the plume of well i. This can be constructed by considering the wells j within the extent of the plume defined by the mass $M^l_{\alpha_i}$. A reasonable model for the draining set is then that $j \in \mathcal{D}^l_{\alpha,i}$ if the plume of well i is the one with the highest saturation at $\boldsymbol{x}_j$. This is a simple rule that deals with overlapping CO_2 plumes, which is a challenge for the semi-analytical modeling framework.

The issue of drainage sets and mass accounting is of greatest interest for the CO_2 phase. However, the transport of brine from deep aquifers to shallower regions may also be of interest, and then also the brine masses must be treated analogously.

Time-Stepping

Systems with both CO_2 and brine require a time-stepping procedure to capture the evolving CO_2 plumes and the complex flow paths of leaking CO_2 and brine. We therefore build upon the pure brine model, and exploit the volume preserving approximation to the convolution integral. At every time-step, a saturation model also needs to be solved.

The general framework follows an IMPES-like algorithm, where the equations are split into a pressure equation and a saturation equation. When solving each of these equations, we can assume the solution of the other is known.

We note that even though we update masses in an explicit fashion, there is no time-step restriction on the saturation update, since there are no finite volume grid blocks. Indeed, the aquifers are essentially grid-free, while the coupled upconing and Darcy-flow in the wells amounts to an implicit solve for the well segments, which in our model have essentially zero volume. The lack of time-step constraint is born out in implementations, as we see very low sensitivity to time-step size.

6.5 APPLICATIONS

We will focus our attention on two applications. The first is a relatively simple system, consisting of brine injection into a system of two permeable aquifers separated and confined by impermeable aquitards. We will consider injection of brine, in the presence of one leaky well. This example will serve to explore in the most transparent setting the three solution strategies proposed in Section 6.4.1.

Our second application explores the more complex setting of CO_2 injection into a geological sequence of formations in Alberta, Canada. This problem concerns not only multiple permeable formations, but also considers a large number of potentially leaking wells, and explores the impact of uncertain parameters on model predictions.

6.5.1 Injection of Brine Near a Leaking Well

The problem of constant-rate brine injection was studied in Section 6.4.1. We recall that pressure buildup was governed by Equations (6.7) and (6.9), which form a closed system of equations when evaluated at every abandoned well. In this section, we will only consider a single leaking well, and assume that all fluid and rock properties are constant and isotropic. Furthermore, for simplicity, we assume that we have two permeable formations which have the same permeability and thickness. This allows us to omit superscripts for parameters, simplifying the exposition. Using the shorthand notation $\Delta_w P^l = P^l(\boldsymbol{x}_1, t) - P^l(\boldsymbol{x}_1, 0)$ to denote the unknown pressure change at the abandoned well in layer l, where $l = 1$ for the lower aquifer and $l = 2$ for the upper aquifer, we can write Equations (6.7) and (6.9) for this problem as

$$\Delta_w P^l = Q_0^l \frac{W(\chi_0(|x_1|, t))}{4\pi\Lambda K} + \frac{\left(\dfrac{dQ_1^l}{dt}\right) * W(\chi_1(r_w, t))}{4\pi\Lambda K}.$$

Here we have assumed that the location of the injection well is at the origin, $\boldsymbol{x}_0 = \mathbf{0}$. In terms of the pressure change, if the initial condition is hydrostatic, the flux can be written as

$$Q_1^l = -(-1)^l \Lambda \bar{K} \frac{\Delta_3 \Delta_w P^l}{h}.$$

Here $\bar{K} = \bar{K}_1^{1+1/2}$ and $h = h^{1+1/2}$. These equations can be combined to yield

$$\Delta_w P^l = Q_0^l \frac{W(\chi_0(|\boldsymbol{x}_1|, t))}{4\pi\Lambda K} - (-1)^l \frac{\bar{K}}{4\pi h K}\left(\frac{d}{dt}\Delta_3 \Delta_w P^l\right) * W(\chi_1(r_w, t)). \quad (6.11)$$

This system, for $l = 1, 2$, can be solved to desired precision by a numerical time-stepping or by numerical inversion of the Laplace transform solution. Additionally, we discussed two simplifications in Section 6.4.1. The first simplification is the volume-preserving approximation which defines a modified time. This simplifies Equation (6.11) from a convolution equation to an integral equation

$$\Delta_w P^l = Q_0^l \frac{W(\chi_0(|x_1|, t))}{4\pi\Lambda K} - (-1)^l \frac{\bar{K}\Delta_3 \Delta_w P^l}{4\pi h K} W\left(\chi_1\left(r_w, \frac{\int_0^t \Delta_3 \Delta_w P^l dt'}{\Delta_3 \Delta_w P^l}\right)\right). \quad (6.12)$$

The second simplification assumes a priori some self-similar scaling of the solution, such that Equation (6.11) can be written without explicit knowledge of the time-history of the problem,

$$\Delta_w P^l = Q_0^l \frac{W(\chi_0(|\boldsymbol{x}_1|, t))}{4\pi \Lambda K} - (-1)^l \frac{\bar{K}\Delta_3 \Delta_w P^l}{4\pi h K} W(\chi_1(r_w, 0.92t)). \qquad (6.13)$$

This last equation is simply a two-by-two linear system of equations for the unknowns $\Delta_w P^l$, *at any time*. This implies that we do not need to consider any time-stepping in order to solve Equation (6.13).

For illustration, we will apply Equations (6.11), (6.12) and (6.13) to a model problem based on representative data. We consider a domain with two flat aquifers of infinite horizontal extent, connected by an abandoned well, where the aquifers are of 30 m thickness, with a permeability of 20 mD. They are separated by 100 m of impermeable aquitard. The leaky well is situated 100 m from the injection well, with an effective radius of 0.15 m and an effective, leaky, permeability of 1 D. The brine modeled with a viscosity of $2.535 \cdot 10^{-4}$ Pa.s, and we consider a compressibility of $4 \cdot 10^{-10}$ Pa^{-1} for this example. Injection takes place for 1000 days.

In Figure 6.2, the three approaches discussed in Section 6.4.1 give indistinguishable results at the resolution of the figure, with all of them falling along the line labeled "converged numerical solution." Also shown in the figure is the leakage response calculated from Equation (2.50), using an aquitard permeability of $5.6 \cdot 10^{-6}$ mD. This illustrates the very different qualitative behavior between leakage through a leaking well, and diffuse caprock leakage. Most importantly, while the injection pressure is the driving force for leakage from the well (as can be seen from the continuous rise of the curves), the actual magnitude is governed by the interaction between the well permeability and the local drawdown immediately in the vicinity of the well. For high-permeable wells, this local drawdown will completely dominate the calculation. In contrast, with the diffuse leakage through the caprock, local draw-down effects are negligible, and the solution is much closer to being proportional to the driving force from the injection well. Keep in mind that in terms of dimensionless parameters, we are looking at relatively early time, and both the leaky aquifer solution and the leaky well solution are still far from their asymptotic values of 1 and 0.5, respectively.

In the context of this example we will also discuss the relative practical merits of Equations (6.11), (6.12) and (6.13). In terms of accuracy, only Equations (6.11) allows for a numerical implementation which is convergent to the exact solution. However, the time-steps of the integration will be limited by the physics, and the convolution integral implies that the computational cost to march to the final time is proportional to the number of time-steps cubed. In contrast, an implementation of Equation (6.12) has a computational cost that is simply proportional to the number of time-steps, while Equation (6.13) can be solved for any time, at a fixed computational cost. However, while the approximations used to derive Equations (6.12) and (6.13) seem applicable for the constant-rate injection shown here, their suitability may deteriorate in the presence of additional wells or more complex injection operations. This problem therefore serves as a transparent example of the value judgments which need to be made when balancing the considerations of accuracy and robustness with computational constrains (see also Box 6.1).

BOX 6.1 *Comparison of Leakage Predictions from CO_2 Storage*

The data in this section is not chosen completely arbitrary. Indeed, this setup was used in early studies for CO_2 leakage from a storage formation, and was also the basis for the first problem in the code intercomparison study we got to know in Box 5.2. For the code comparison, the data from the text was supplemented by the fluid properties of the deep, warm formations of Table 3.1, and a square finite outer boundary was imposed with side-length 1 km.

We see the results from the 13 participating groups in Figure 6.A. It is interesting to note that despite the relative simplicity of the problem, there is still not a convergence of the results. The methods presented in this book were used by the entries *VESA (Princeton Uni)*, which used numerical implementation as described in Section 6.3, and The *Semi-anal. sol., ELSA (Uni Bergen/Princeton Uni)*, which was an implementation following Section 6.4. Both implementations used the *Saturation model for leaking wells* described in Section 6.2.

All other results in the figure are from three-dimensional numerical codes, and as such should in theory be convergent to the true solution of the governing equations, if a fine enough grid is used. The results with TOUGH2 from Melbourne highlight the challenges of three-dimensional simulation. They reported fine-grid solutions for early times that differed noticeably from the coarser-grid solutions; however, the computational cost did not allow for a fine-grid simulation over the whole time period. As we saw in Box 5.2, the issue of computational cost versus resolution has an even more pronounced impact for more complicated parameter fields.

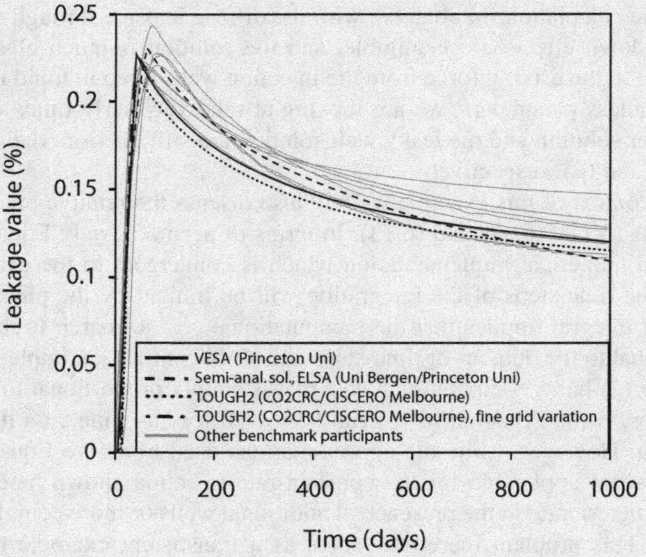

Figure 6.A Results from problem 1 of the Stuttgart CO_2 storage code intercomparison project (modified from Class et al., 2009).

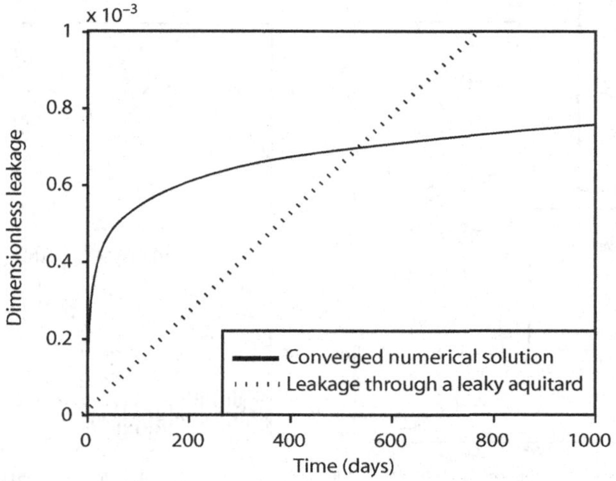

Figure 6.2 Leakage as a function of time for a single-phase brine system.

6.5.2 Injection of CO_2 in a Realistic Geological Setting

This application will use the concepts presented in this chapter to model potential leakage of both CO_2 and brine, using hypothetical but realistic injection operations in the Alberta Basin. This large basin in western Canada has a number of permeable formations within a stratigraphic sequence dominated by alternating permeable and impermeable layers. To illustrate the kinds of simulations that can be performed, we focus on the Wabamun Lake area, which is shown in Figure 6.3a. Among other things, this area has four large coal-fired power plants in its vicinity, emitting collectively about 30 million metric tonnes of CO_2 per year. So this seems like a logical possibility for a large-scale carbon capture and storage (CCS) operation.

Figure 6.3 shows the locations of all existing oil and gas wells in the Wabamun Lake area, over a domain that is 50 km by 50 km. There are about 1,250 wells in the area, giving a density of about 0.5 wells per square kilometer. This is not an unusual density for this part of the world, see Figure 1.7. If we assume that the CO_2 plume size coupled with the lateral extent of pressure perturbations will cover much of this domain, then we see clearly that the potential for leakage along wells may be significant and would need to be considered in any comprehensive risk assessment associated with a CCS operation.

Figure 6.3 also shows the stratigraphic sequence along with the depth of all of the wells in the domain, with all wells projected along a single east-west transect. This figure reminds us that the depth of penetration of wells (and any other potential leakage conduits) is important, and makes the three-dimensionality of the real problem very clear. Within the modeling framework developed herein, it is important

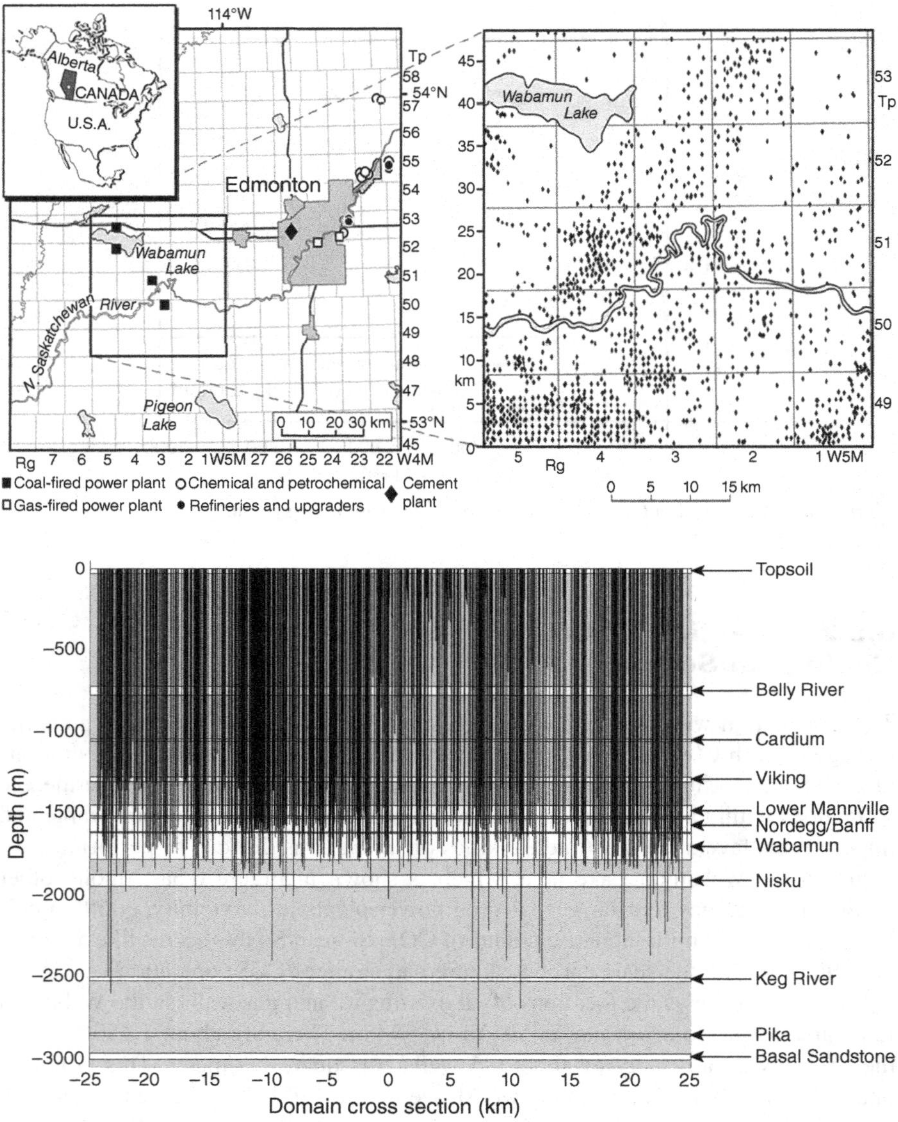

Figure 6.3 Upper left figure pane shows the study site southwest of Edmonton, Alberta. Black boxes designate the location of large point sources of CO_2. In the upper right pane, all known active and abandoned wells within the study are marked by dots. The depth of the wells, projected onto an east-west transect, are shown in the lower pane.

to know in which formation a given well has been completed. Simple inspection of Figure 6.3 would indicate that risk of leakage decreases with depth, simply because fewer wells continue to deeper depths. In the Basal Sandstone Formation, only one well penetrates through its overlying caprock, so risk of leakage along old wells would be minimal for injection into that deep formation. Compare this to, for example, the Viking Formation, which has about 900 wells penetrating through its overlying caprock formation within the same 50 km × 50 km domain.

To illustrate the utility of the modeling tools developed in this chapter, we consider the problem of injection of CO_2 into one of the formations in the stratigraphic sequence, and the subsequent spread of both the pressure pulse and the CO_2 plume together with leakage of CO_2 and brine from the injection formation to other formations via leaky wells. Before presenting calculations, let us consider the overall behavior of the system, and the kinds of data required to perform these simulations.

In order to model this system, a significant amount of data is required. In addition to all of the parameters needed for the injection formation (permeability, porosity, thickness, relative permeability, capillary pressure, compressibility, etc.), similar sets of parameters and functional relationships are needed for all of the other permeable layers within which secondary plumes may form. Furthermore, a set of properties describing the hydraulic characteristics of the materials that comprise a "leaky well" must be specified. If the model of Equation (6.4) is used along wells, then the major parameter is the effective permeability of the leaky well and the effective cross-sectional area over which flow takes place. While all of these parameters are uncertain, and may have complex patterns of spatial variability, the most uncertain of these parameters will almost always be those associated with the wells. This motivates a probabilistic analysis of the system, with well permeability as the variable (or stochastic) parameter. The need for probabilistic modeling, in which many thousands of simulations are performed in order to sample properly the distributions of input variables, is a strong motivation for simplified models like the semi-analytical approach described in this chapter.

Based on studies performed by the Alberta Geological Survey, properties for each of the formations in the stratigraphic sequence can be estimated. A few of the key parameters for these formations are presented in Table 6.1, including the formation thickness, formation (average) permeability, and the number of wells perforating the caprock formation overlying the formation. The highlighted formations—Nordegg/Banff, Nisku, and Basal Sandstone—are specific formations into which injection has been simulated in earlier and ongoing studies.

We present results for injection into the Nisku formation using an assumed end-point relative permeability for the CO_2 of 0.55 with injection at the maximum rate dictated by fracture pressure constraints. We perform simulations using 50 years of continuous injection and results are presented at the end of the 50 years of injection. We apply the semi-analytical approximation described in Section 6.4, using the volume-preserving approximation to simplify the convolution integrals. The overall system is 50 km by 50 km, with 10 permeable layers and 1,237 wells. The data correspond to the information presented in the earlier figures and tables. While

Table 6.1 Properties of Permeable Formations in Study Area

Aquifer name*	Top depth [m]	Thickness [m]*	Intrinsic permeability [miliDarcy]*	Wells ending in layer
Belly River	728.8	56	86	1237
Cardium	1051.8	15	7	1155
Viking	1287.8	30	53	900
Mannville	1461.8	65	7	895
Nordegg/Banff	1537.8	80	4	733
Wabamun	1628.8	160	4	138
Nisku	1881.8	72	170	39
Keg River	2506.8	22	3.5	11
Pika	2844.8	14	16	2
Basal Sandstone	2964.8	38	23	1

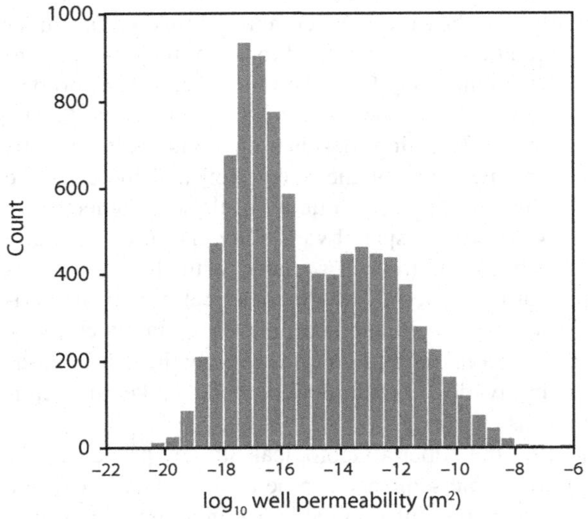

Figure 6.4 Modeled distribution of well-segment effective permeability.

simulation of this kind of system using traditional numerical simulators is computationally prohibitive, we can perform a complete 50-year simulation in minutes of computing time on a single processor. This allows us to perform many simulations and to look at both the input and the output in a probabilistic framework.

We define the effective permeability of each individual well segment based on a bimodal lognormal distribution of permeability values like that shown in Figure 6.4. Each permeability value is assigned randomly, based on this distribution. The values are either assigned independent of all others (no correlation), or they are assigned so that the property is the same for all segments in a given well (correlation along the well). For these two different correlation structures, we run 1,000 simula-

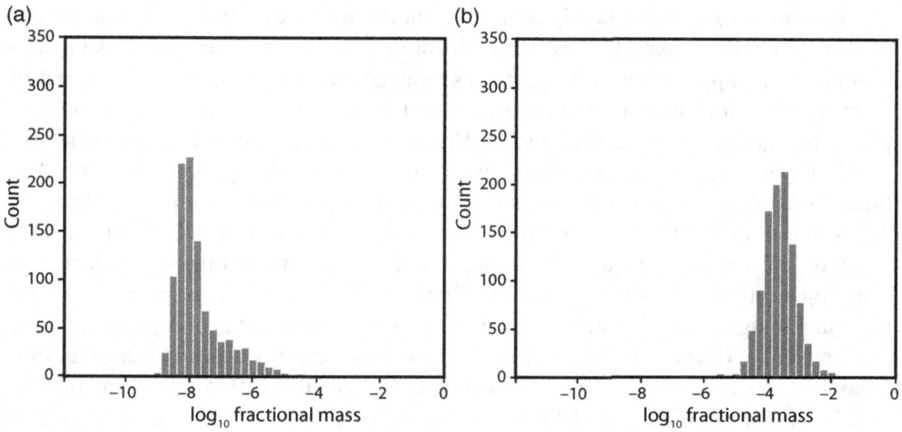

Figure 6.5 Leakage of CO_2 into uppermost modeled formation, Belly River. Representative as an upper bound on estimates of leakage out of the system, for example, back to the atmosphere. Part (a) corresponds to each vertical well segment having an independently sampled random effective permeability, while (b) corresponds to each well having a single value of effective permeability throughout its vertical extent.

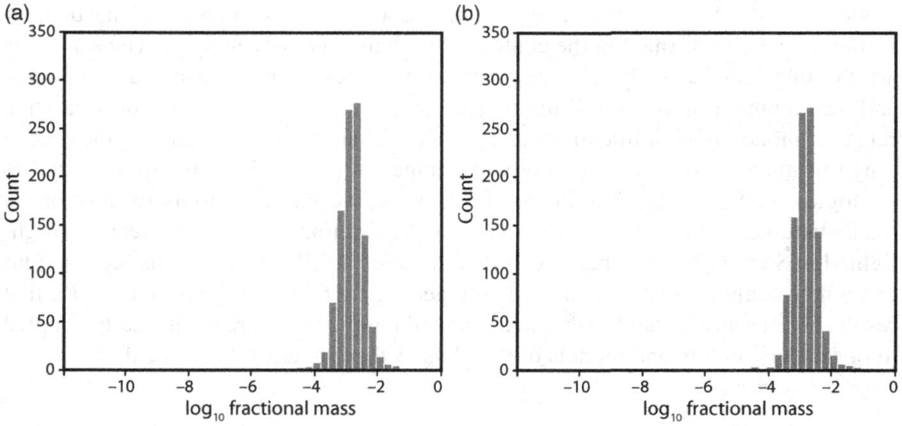

Figure 6.6 Leakage of CO_2 into the Wabamun formation, which is the first formation above Nisku. Representative of leakage out of the injection formation. Part (a) corresponds to each vertical well segment having an independently sampled random effective permeability, while (b) corresponds to each well having a single value of effective permeability throughout its vertical extent.

tions, each using a different set of assigned well permeabilities but using the same distribution from which the permeability values are chosen. We then calculate, from all of the output, the amount of each fluid—CO_2 and brine—that has leaked out of the injection formation via well leakage. This includes the total amount of CO_2 and brine that leaks from the injection formation, independent of where it goes. These results are shown in Figures 6.5 and 6.6.

Two observations can be highlighted from these figures. The first is that the correlation structure makes a very large difference in the amounts of leakage that migrates to the upper layer of the vertical sequence. The second is that the correlation structure does not make a large difference in terms of the total amount of CO_2 that leaves the injection formation. These results can be generalized in terms of the impact of correlation structure on the amount of CO_2 that reaches the uppermost layers. One-dimensional flow in the wells means that any low-permeability segment will greatly reduce the overall flow rate, and therefore uncorrelated values will be much more effective at stopping leakage. It is important to remember this when designing sampling strategies for well properties. More generally, these broad results can give important insights into the underlying dynamics in these systems of injection, migration, and leakage. They also provide the tools to explore many other aspects of system behaviors. For example, while not shown here, we see quite different leakage profiles for CO_2 and brine, with more CO_2 leaking upward because it is driven by both pressure and buoyancy while brine leakage is driven only by pressure.

The models presented in this chapter capture the dominant physical processes in the system, including two-phase flows driven by both pressure gradients and buoyancy within formations and along leakage pathways. The models also include the dominant geometries including complex large-scale geological features and large numbers of critical small-scale leakage pathways. A remarkable aspect of these models is that they are extremely efficient, so that many thousands of stochastic realizations can be performed in the context of probabilistic risk analysis. These models arise through careful analysis of length and time scales, judicious use of assumptions, and development of an overall multi-scale modeling framework that allows a wide range of modeling simplifications and choices. That framework can provide a wide range of models for concomitantly wide ranges of practical calculations related to geological CO_2 storage. Within that framework, we have the tools to develop an understanding of the system sensitivity to both modeling of physical effects, as highlighted in Section 5.5, and parameter uncertainties, as illustrated in this section. Our example calculations show that modeling geological CO_2 storage is an exercise that results not in a single number, but in a range of numbers that are intrinsically coupled to our understanding and models of the physics that govern subsurface flows.

6.6 FURTHER READING

This chapter has presented the extent of our current understanding of large-scale modeling of CO_2 storage systems. The further reading therefore does not necessarily take the reader further than the content of the chapter, but rather refers to literature where these ideas were first developed.

Papers

Celia, M.-A., J.-M. Nordbotten, S. Bachu, M. Dobossy, and B. Court. 2008. Risk of leakage versus depth of injection in geological storage. Proceedings of the 9th Conference on Greenhouse Gas Technologies, Washington DC, USA, November 16–20.

Class, H., A. Ebigbo, R. Helmig, H. Dahle, J.M. Nordbotten, M.A. Celia, P. Audigane, M. Darcis, J. Ennis-King, Y. Fan, B. Flemisch, S.-E. Gasda, S. Krug, D. Labregere, J. Min, A. Sbai, S.-G. Thomas and L. Trenty. 2009. A benchmark study on problems related to CO_2 storage in geologic formations. Special issue of *Computational Geosciences*, 13(4), 409–434.

Hunt, B. 1985. Flow to a well in a multiaquifer system. *Water Resources Research*, 21, 1637–1641.

Nordbotten, J.-M., M.A. Celia, and S. Bachu. 2004. Analytical solutions for leakage rates through abandoned wells. *Water Resources Research*, 40(4), W04204.

Nordbotten, J.-M., M.-A. Celia, S. Bachu, and H.-K. Dahle. 2005. Semi-analytical solution for CO2 leakage through an abandoned well. *Environmental Science & Technology*, 39(2), 602–611.

Nordbotten, J.-M., D. Kavetski, M.-A. Celia, and S. Bachu. 2009. A semi-analytical model estimating leakage associated with CO2 storage in large-scale multi-layered geological systems with multiple leaky wells. *Environmental Science and Technology*, 43(3), 743–749.

Peaceman, D.W. 1983. Interpretation of well-block pressures in numerical reservoir simulation with nonsquare grid blocks and anisotropic permeability. *Society of Petroleum Engineers Journal*, 23, 531–543.

Data

Alberta Geological Survey (Wabamun data set). http://www.ags.gov.ab.ca/co2_h2s/wabamun/Wabamun_base.html

Epilogue

We come to the end of our journey into the modeling of carbon storage, a journey that has involved some fairly complicated mathematics but also a number of important concepts on which the derivations were based. Even if you have not followed all of the mathematical details, we hope that the modeling philosophies will stick with you. The carbon problem needs practical answers to important practical questions, and these answers must be based on sound physical and mathematical reasoning. In that regard, we have tried to follow a philosophy that matches the mathematical model to the question being asked. We did this with explicit considerations of length and time scales, and development of mathematical models describing the subsurface system at the length and time scales that matched the questions being asked. This necessarily leads to the creation of specific algorithms to allow information to be represented across a range of length and time scales. To accomplish this, we used the general concept of multiscale modeling, which is particularly suited to the carbon problem. In fact, the strong buoyant segregation that characterizes the CO_2–brine system makes multiscale modeling based on vertical equilibrium a very natural approach.

Multiscale modeling was based on the idea of compression of information into simplified representations, and subsequent reconstruction of the compressed values back into their original expanded form. When used judiciously, this concept allows for greatly simplified calculations, even if the complexity of mathematical notation sometimes obscures the simple nature of the equations. Among many other things, we find very useful analytical solutions for simple injection of CO_2 into a deep saline aquifer, including solutions for both the location of the CO_2 and the evolution of the pressure field. These solutions can form the basis for more complex calculations, including leakage along multiple wells over multiple geological formations, as well as other studies that give sensitivity information about system parameters. For more involved calculations, including spatial heterogeneity, irregular geometries, and other complications, simplified numerical calculations can still be achieved with the sole assumption of vertical equilibrium. Of course, the informed reader or analyst will ask: Is vertical equilibrium of fluid pressures a reasonable assumption? The

Geological Storage of CO_2: Modeling Approaches for Large-Scale Simulation, First Edition.
Jan M. Nordbotten and Michael A. Celia.

answer lies in a fairly simple analysis of time scales associated with vertical segregation, as we have tried to explain to our readers. If the time scale of segregation is small relative to the time scale of the problem being modeled, then it is a good assumption. If not, then a full simulation including significant vertical velocities is probably necessary. Our own experience is that vertical equilibrium is often a very good assumption. Answers to similar questions about other possible assumptions follow along the same lines, where analysis of relevant space and time scales gives the proper insights. We believe that this kind of systematic approach to model development is very powerful, allowing complex problems to be analyzed systematically and transparently. This will inevitably lead to better model formulation and more informed applications.

While presenting this particular philosophy—in addition to all of the background materials—we also recognize that we have omitted a number of aspects that we could have included. This includes detailed numerical solution methodologies, details of geochemistry and the wide range of geochemical reactions that could occur in these systems, details of non-isothermal effects and solutions to the energy equation, and coupled models involving combinations of these different types of processes. We were motivated to exclude these detailed discussions, in part, by the fact that the underlying two-phase flow physics and associated pressure responses in the system are often the most important aspects when trying to answer practical questions. We are also very often data-limited when we are making an analysis, especially early on in a project, and lack of data will also drive us toward simpler models from which we can gain the most insights.

In general, we need to accept that the subsurface is a highly uncertain environment, and that models meant to predict fluid behavior in these environments will necessarily also be highly uncertain. With the data available, the mathematical modeler, the engineer, the consultant, and the reader of this book should be able to use appropriate mathematical models to give a "best estimate" on how a CO_2 storage operation will evolve. The analyst can highlight which physical effects are expected to dominate, and discuss the relative merits of different strategies. In this way, an extensive list of "what if" scenarios can be investigated, thereby shaping an understanding of the interplay between uncertainty and risk. In short, through modeling we can make a difference by ensuring that CO_2 storage is conducted more safely, more efficiently, and with a greater respect for the challenges and inherent uncertainties. We hope that the approaches and techniques described in this book can help in this overall process.

Appendix

A1 GENERAL CONVENTIONS

Where two scales are discussed simultaneously, we try to be consistent using lowercase letters to denote variables defined on the finer scale, and uppercase letters to denote variables defined on the coarser scale. Variables that have the same interpretation at both scales, are denoted by the same letter, for example, permeability is denoted by k on the finer scale and K on the coarser scale. The use of transmissivity T for an integrated hydraulic conductivity κ forms an exception. Note that variables may change units with scale.

We use the convention that vectors and matrices are indicated by a bold font. We do not use the convention of summation over repeated indexes.

We often recycle symbols for variables with similar function or interpretation. In these cases, the variables are distinguished by different sub- or superscripts. Note that the units of these variables may be different.

As far as possible, we have tried to use the conventional symbols for variables where there is a well established convention. Where this leads to confusion, due to multiple variables having the same symbol in different fields of study, we try to distinguish variables by using scripted fonts. However, some letters (such as T) are so common that we have been unable to give all its different uses different visual expression. In these cases, we have tried to ensure that the correct interpretation be clear from the context.

A2 ARABIC LETTERS

Symbol	Description
a	As subscript: air
A	Area
b	As subscript: brine

(Continued)

Geological Storage of CO_2: Modeling Approaches for Large-Scale Simulation, First Edition.
Jan M. Nordbotten and Michael A. Celia.

Symbol	Description
B	As subscript: Bottom
c	Context dependent coefficient
c	Fine scale compressibility
c	As subscript: CO_2
C	Context dependent coefficient
C	Coarse scale compressibility
$\mathcal{C}$	Compression operator (fine to coarse scale)
d	As subscript: drainable
d	Ordinary derivative
∂	Boundary
∂	Partial derivative
D	As superscript: Dupuit
D	Total derivative
$\boldsymbol{e}$	Unit vector
e	As subscript: Energy
E	Specific energy content
E	Energy dimension
f	Fine scale mass flux
f	Fractional flow function
f	As subscript: fluid
$\mathscr{f}$	Generic conserved quantity flux vector
F	Coarse-scale mass flux
F	Force dimension
g	Gravitational constant
$\boldsymbol{g}$	Gravitational vector
G	As subscript: Global
$\boldsymbol{G}$	Coarse gravitational variable
h	Height, usually of water column
h	Hydraulic head
$\mathscr{h}$	Specific enthalpy
H	Aquifer thickness
i	Summation index
i	Component index
I	As subscript: Interface
I	As subscript: Infiltration
$\boldsymbol{j}$	Non-advective mass flux
$\boldsymbol{j}_e$	Non-advective energy flux
j	Summation index
J	Bessel function of the first kind
k	Small-scale permeability
K	Large-scale permeability
K	Modified Bessel function of the second kind
$\mathcal{K}$	Thermal conductivity
l	Length of pore throat

Symbol	Description
ℓ	Length (of column in Darcy experiment)
ℓ	Length scale
L	Length dimension
L	Length of network model
m_{α}^{i}	Mass fraction of component i in phase α
$\mathscr{m}$	Generic conserved quantity per volume
$\mathcal{M}$	Mass dimension
M	Coarse mass density
M	As subscript: Mobile
n	Time step in numerical discretization
N	Number of time steps
N	Upper limit of summation
N	Dimension
p	Fine-scale pressure
P	Coarse-scale pressure
q	Fine-scale volumetric flow rate
Q	Coarse-scale volumetric flow rate
r	Radial coordinate or radius
r	As subscript: Relative
$\mathscr{r}$	Generic conserved quantity source per volume
$\mathcal{R}$	Reconstruction (coarse to fine)
R	As subscript: Residual
s	Fine-scale saturation
s	As subscript: solid
s	As subscript: specific
S	Storativity (specific when subscript s)
t	Temporal coordinate
t	As superscript: Time history of variable
T	Time dimension
T	Transmissivity
T	As subscript: Top
T	As superscript: Transpose
$\boldsymbol{u}$	Fine-scale fluid flux
$\boldsymbol{U}$	Coarse-scale flux
$\boldsymbol{v}$	Velocity
$\boldsymbol{\mathscr{v}}$	Fluid velocity
V	Volume
w	As subscript: wetting
W	Well function
x	Spatial coordinate
x_{α}^{i}	Molar fraction of component i in phase α
y	As subscript: yield
z	Vertical coordinate

A3 GREEK LETTERS

Symbol	Description
α	Phases index (usually subscript)
β	Phases index (usually subscript)
γ	Interfacial tension
Γ	Dimensionless group
δ	Dirac delta distribution
Δ_α	Difference operator between phase properties
$\Delta_{x_1}^{i,j}$	Spatial central difference operator
Δ^*	Difference across shock/interface
Δ_3	Difference between geological layers
ϵ	Small number
ζ	Surface
η	Dimensionless space coordinate
θ	Angle (coordinate)
ϑ	Basis function
Θ	Temperature
κ	Fine scale hydraulic conductivity
λ	Fine-scale mobility
Λ	Coarse-scale mobility
μ	Dynamic viscosity
ν	Normal vector
ξ	Fine-scale solution vector
Ξ	Coarse-scale solution vector
π	Ratio of circumference to diameter of circle
Π	Product
ρ	Fine-scale density
ϱ	Coarse-scale density
σ	Effective stress
Σ	Summation
τ	Dimensionless time coordinate
υ	Fine-scale volumetric source density
Υ	Coarse-scale volumetric source density
ϕ	Fine-scale porosity
Φ	Coarse-scale porosity
χ	Self-similar coordinate
ψ	Fine-scale mass source density
Ψ_h	Fine-scale enthalpic source density
Ψ	Coarse-scale mass source density
ω	Domain, usually part of Ω
Ω	Domain, usually whole domain of interest

A4 MULTI-LETTER NOTATION

Letters	Description
nw	As subscript: nonwetting
cap	As superscript: Capillary
MW	Molecular weight
res	Residual
eff	As subscript: Effective
th	As subscript: Thermal
int	As subscript: Interface
lab	As subscript: Laboratory
II	As superscript: Immiscible and incompressible
SI	As superscript: Sharp Interface
VS	As superscript: Vertically Structured
CM	As superscript: Convective Mixing
FE	As superscript: Finite Element

A5 NON-LETTER SYMBOLS

In this section, we use u and v as arbitrary functions or expressions.

Notation	Description	Definition or usage
$[u]$	Brackets indicate that the subject of the sentence has dimension u	for example, $[L^3]$
∇	Nabla is the vector derivative operator	$\nabla \equiv e_1 \frac{\partial}{\partial x_1} + e_2 \frac{\partial}{\partial x_2} + e_3 \frac{\partial}{\partial x_3}$
$\parallel$	Parallel lines indicate the two first coordinate directions	$\nabla_{\parallel} \equiv e_1 \frac{\partial}{\partial x_1} + e_2 \frac{\partial}{\partial x_2}$; $u_{\parallel} = e_{\parallel} \cdot u = u_1 e_1 + u_2 e_2$
$\bar{u}$	Overbar denotes vertical average of u	
$\tilde{u}$	Tilde denotes variation from average u	
$\hat{u}$	Hat denotes a reconstructed variable	$\hat{u} = \mathcal{R}U$
*	Star as superscript refers to shock	s^*
*	Star as subscript refers to critical dimensionless distance	χ_*
$\lvert u \rvert$	Lines indicate absolute value, or length of a vector	$\lvert \boldsymbol{u} \rvert \equiv \sqrt{\boldsymbol{u} \cdot \boldsymbol{u}}$
u'	Prime refers to endpoint of mobility curve	$\lambda'_\alpha = \lambda_\alpha(s_\alpha = 1)$
$\underline{u}$	Underbar denotes a space of functions	
(u_1, u_2)	Parenthesis with comma denotes function inner product	$(u_1, u_2) \equiv \int_\Omega u_1 \cdot u_2 \, d\boldsymbol{x}$
$[u_1, u_2]$	Brackets with comma denotes function inner product in space-time	$[u_1, u_2] \equiv \int_0^\infty \int_\Omega u_1 \cdot u_2 \, d\boldsymbol{x} \, dt$
$[u_1; \ldots ; u_N]$	Brackets with colon (or comma) denote the definition of a vector.	

A6 VECTOR NOTATION

This section serves as a brief review and reference for readers more familiar with index or tensor notation.

A6.1 Vectors and Matrices

We denote vectors and matrices by bold face symbols. The unit (column) vectors are denoted $\boldsymbol{e}_i$, and are defined such that element i of the vector is equal to 1, while the remaining elements are zero. These vectors form the basis of a Cartesian coordinate system. Thus, all vectors can be decomposed as $\boldsymbol{u} = \sum_{i=1}^{N} \boldsymbol{e}_i u_i$. To specify the elements of a column vector we can also write $\boldsymbol{u} = [u_1; \ldots ;u_N]$, where the elements are separated by semi-colons. We denote a row vector by the transpose of a column vector, $\boldsymbol{u}^{\mathrm{T}}$. To specify the elements of a row vector we separate the elements by commas, and write $\boldsymbol{u} = [u_1, \ldots, u_N] = [u_1; \ldots; u_N]^{\mathrm{T}}$. A matrix $\boldsymbol{A}$ can be represented as a (column) vector of row vectors, $\boldsymbol{A} = \left[\boldsymbol{a}_1^{\mathrm{T}}; \ldots; \boldsymbol{a}_N^{\mathrm{T}}\right] = \sum_{i=1}^{N} \boldsymbol{e}_i \boldsymbol{a}_i^{\mathrm{T}}$. In this appendix we will for clarity use capital letters for matrices; however, in the remainder of the book capitalization bears no relation to whether the variable is a vector or matrix.

In the last definition we used the product of two vectors. Vector multiplication follows the usual conventions for multiplication of matrices. Thus, the product $\boldsymbol{u}^{\mathrm{T}}\boldsymbol{v} = \sum_i^N u_i v_i$ is a scalar, while $\boldsymbol{u}\,\boldsymbol{v}^{\mathrm{T}} = [u_1\boldsymbol{v}^{\mathrm{T}}; \ldots ;u_N\boldsymbol{v}^{\mathrm{T}}]$ is a matrix. The former also forms the definition of the inner product (or so-called dot product), $\boldsymbol{u}\cdot\boldsymbol{v} \equiv \boldsymbol{u}^{\mathrm{T}}\boldsymbol{v} = \sum_{i=1}^{N} u_i v_i$.

The products between matrices is defined directly from the vector products. Thus, for a vector $\boldsymbol{u}$ and matrices $\boldsymbol{A}$ and $\boldsymbol{B}$, we have the following products. A matrix times a vector forms a vector of vector inner products:

$$\boldsymbol{A}\boldsymbol{u} = \left[\boldsymbol{a}_1^{\mathrm{T}}; \ldots; \boldsymbol{a}_N^{\mathrm{T}}\right]\boldsymbol{u} = \left[\boldsymbol{a}_1^{\mathrm{T}}\boldsymbol{u}\,; \ldots; \boldsymbol{a}_N^{\mathrm{T}}\boldsymbol{u}\right]$$

A transposed vector times a matrix is defined by transpositions:

$$\boldsymbol{u}^{\mathrm{T}}\boldsymbol{A} = \left(\boldsymbol{A}^{\mathrm{T}}\boldsymbol{u}\right)^{\mathrm{T}}$$

A matrix times a matrix forms a new matrix defined as a vector of vector-matrix products:

$$\boldsymbol{A}\boldsymbol{B} = \left[\boldsymbol{a}_1^{\mathrm{T}}; \ldots; \boldsymbol{a}_N^{\mathrm{T}}\right]\boldsymbol{B} = \left[\boldsymbol{a}_1^{\mathrm{T}}\boldsymbol{B}; \ldots; \boldsymbol{a}_N^{\mathrm{T}}\boldsymbol{B}\right]$$

We generalize the notation of the dot product of two vectors to imply transposition of the first argument, thus the dot product between a vector and a matrix is defined as

$$\boldsymbol{u}\cdot\boldsymbol{A} \equiv \boldsymbol{u}^{\mathrm{T}}\boldsymbol{A}$$

This defines the vector and matrix multiplications needed for the majority of this book. Additionally, we will briefly need the cross product, which defines the vector $\boldsymbol{c} = \boldsymbol{a} \times \boldsymbol{b}$. The cross product has the property that the resulting vector $\boldsymbol{c}$ is orthogonal to both the arguments $\boldsymbol{a}$ and $\boldsymbol{b}$, with a length equal to the area of a parallelogram which sides are given by $\boldsymbol{a}$ and $\boldsymbol{b}$.

A6.2 Inverse Matrix

The inverse matrix is defined using the concept of the identity matrix, denoted by $\boldsymbol{I}$. Let the identity matrix be defined as $\boldsymbol{I} = \sum_{i=1}^{N} \boldsymbol{e}_i \boldsymbol{e}_i^{\mathrm{T}}$. The identity matrix has the property that $\boldsymbol{Iu} = \boldsymbol{u}$ for all vectors $\boldsymbol{u}$. For a matrix $\boldsymbol{A}$, we denote its inverse (when it exists) as $\boldsymbol{A}^{-1}$, and define it as the unique matrix that has the property that $\boldsymbol{AA}^{-1} = \boldsymbol{A}^{-1}\boldsymbol{A} = \boldsymbol{I}$.

A6.3 Vector Differential Operators

We define the vector differential operator in 3D as

$$\nabla \equiv \boldsymbol{e}_1 \frac{\partial}{\partial x_1} + \boldsymbol{e}_2 \frac{\partial}{\partial x_2} + \boldsymbol{e}_3 \frac{\partial}{\partial x_3}$$

When this operator is applied to a scalar, say h, it is referred to as the "gradient" operator and written in Cartesian coordinates as

$$\nabla h = \frac{\partial h}{\partial x_1} \boldsymbol{e}_1 + \frac{\partial h}{\partial x_2} \boldsymbol{e}_2 + \frac{\partial h}{\partial x_3} \boldsymbol{e}_3$$

Similarly, we can apply the vector differential operator to a vector function, which is referred to as the "divergence" operator and defined in 3D as

$$\nabla \cdot \boldsymbol{u} = \nabla^{\mathrm{T}} \boldsymbol{u} = \frac{\partial u_1}{\partial x_1} + \frac{\partial u_2}{\partial x_2} + \frac{\partial u_3}{\partial x_3}$$

Applying both the gradient and the divergence in succession gives the so-called *Laplacian Operator*, which is denoted (slightly imprecisely) as $\nabla^2 = \nabla \cdot \nabla$. The Laplacian applied to a scalar can be written out in 3D as

$$\nabla^2 h = \nabla \cdot (\nabla h) = \frac{\partial^2 h}{\partial x_1^2} + \frac{\partial^2 h}{\partial x_2^2} + \frac{\partial^2 h}{\partial x_3^2}$$

While the theory of differential operators on vectors has a beautiful generalization, the divergence, gradient, and Laplacian operators will be sufficient in this book. It is useful to think of these operators not in terms of their definitions, but rather in terms of their physical implications: the gradient operator gives the gradient of a multivariate scalar function, and the divergence operator gives the divergence of a velocity field. Using vector notation to simplify the visual impression of the equations should encourage the reader to think in terms of the physical interpretation of the differential operators.

A6.4 Gauss and Leibnitz

Consider a vector field $\boldsymbol{u}$, which represents the flow of fluid. If we want to know the amount of fluid flowing out of a domain we need the component normal to the boundary, $\boldsymbol{u} \cdot \mathbf{v}_n$, where $\mathbf{v}_n$ is the normal vector to the boundary, pointing outward. Similarly, the loss of fluid at any point is given by the divergence of $\boldsymbol{u}$. Intuitively, we expect that the total flow out of the domain is equal to the total loss of fluid at all points in the domain. The mathematical statement and proof of this is known as Gauss' theorem, which in our notation is stated as

$$\oint_{\partial\Omega} \boldsymbol{F} \cdot \boldsymbol{v}_n dA = \int_{\Omega} \nabla \cdot \boldsymbol{F} dV$$

This is a special case of the more general Stokes' theorem, both of which in 1D simplify to the Fundamental Theorem of Calculus.

Leibnitz's rule applies to the derivative of integrals. In the simplest case, we consider the function $h(x,y)$, which is integrated with respect to its first variable over a domain bounded by two functions $f_1(x)$ and $f_2(x)$. Leibnitz's rule now formalizes the intuitive understanding that the derivative of this integral is dependent on the variation both of the integrated function h and the integration limits:

$$\frac{d}{dx}\int_{f_1(x)}^{f_2(x)} h(x,y)dy = h(x, f_2(x))\frac{df_2}{dx} - h(x, f_1(x))\frac{df_1}{dx} + \int_{f_1(x)}^{f_2(x)} \frac{\partial}{\partial x} h(x,y)dy$$

In the case where $h = h(x_1)$, the last term is zero such that Leibnitz's rule also simplifies to the Fundamental Theorem of Calculus. Leibnitz's rule extends in the natural way to the gradient and divergence operators:

$$\nabla\int_{f_1(x)}^{f_2(x)} h(\boldsymbol{x},y)dy = h(\boldsymbol{x}, f_2(\boldsymbol{x}))\nabla f_2 - h(\boldsymbol{x}, f_1(\boldsymbol{x}))\nabla f_1 + \int_{f_1(x)}^{f_2(x)} \nabla h(\boldsymbol{x},y)dy$$

$$\nabla\cdot\int_{f_1(x)}^{f_2(x)} \boldsymbol{u}(\boldsymbol{x},y)dy = \boldsymbol{u}(\boldsymbol{x}, f_2(\boldsymbol{x}))\cdot\nabla f_2 - \boldsymbol{u}(\boldsymbol{x}, f_1(\boldsymbol{x}))\cdot\nabla f_1 + \int_{f_1(x)}^{f_2(x)} \nabla\cdot \boldsymbol{u}(\boldsymbol{x},y)dy$$

A6.5 Cylinder Coordinates

For easy reference we also give the vector differential operators in cylinder coordinates:

$$\nabla h = \frac{\partial h}{\partial r}\boldsymbol{e}_r + \frac{1}{r}\frac{\partial h}{\partial \theta}\boldsymbol{e}_\theta + \frac{\partial h}{\partial z}\boldsymbol{e}_z$$

$$\nabla \cdot \boldsymbol{u} = \frac{1}{r}\frac{\partial}{\partial r}(ru_r) + \frac{1}{r}\frac{\partial u_\theta}{\partial \theta} + \frac{\partial u_z}{\partial z}$$

For radial coordinates in 2D, simply recall that the functions have no vertical dependence, so that all $\partial/\partial z$ terms are zero.

A6.6 Dirac delta distribution

The Dirac delta distribution for a scalar argument is a distribution (a function i.e., only defined when integrated) that has the following properties:

1. The Dirac delta is zero everywhere but the origin; $\delta(x) = 0$ if $x \neq 0$.
2. The integral of the Dirac delta is equal to one; $\int_{-r}^{r} \delta(x')dx' = 1$ for all $r > 0$.

It is useful to note that from the integral property, the Dirac delta can be understood to have units that are the inverse of the units of its argument (usually time or space).

We will sometimes use the notation of a vector argument for the Dirac delta. The two properties of the Dirac delta are then

1. The Dirac delta is zero everywhere but the origin; $\delta(\boldsymbol{x}) = 0$ if $\boldsymbol{x} \neq 0$.
2. The integral of the Dirac delta is equal to one; $\int_{\omega_r} \delta(\boldsymbol{x}')d\boldsymbol{x}' = 1$ for all $r > 0$.

Here ω is some domain containing the origin and all points within a distance r. The integral property now gives the interpretation that the unit of the Dirac delta is the inverse of the *product* of the units of the argument. For example, if $\boldsymbol{x}$ is a spatial vector in $\mathbb{R}^N$, where each dimension has units [L], then the units of the Dirac delta should be interpreted as $[\mathrm{L}^{-N}]$.

A useful construction of the Dirac delta is to consider a limit of a square pulse functions. Let $P(\boldsymbol{x})$ be defined for such that $P(\boldsymbol{x}) = 1$ if $x_i < 1/2$ (for all components x_i of $\boldsymbol{x}$) and $P(\boldsymbol{x}) = 0$ otherwise. Then the Dirac delta distribution can be defined (in the weak sense of the limit) as

$$\delta(\boldsymbol{x}) = \lim_{\epsilon \to 0} \epsilon^{-N} P(\epsilon^{-1}\boldsymbol{x})$$

This definition clearly satisfies the integral identity (property 2) when $\epsilon << r$, and in the limit also satisfies that the Dirac delta is zero away from the origin (property 1). We note that this is one of many ways to construct a limit that converges to the Dirac delta distribution. Indeed, replacing P with by any positive continuous function with unit integral (such as a normal distribution), will not change the above limit. A particularly useful definition is obtained in one dimension, if the notion of weak derivatives is allowed, where the Dirac delta function can be defined as the (weak) derivative of the Heaviside step function.

The Dirac delta distribution is less useful in radial coordinates, since it is of most relevance to place it at the origin. However, since the origin is a singularity of the domain in radial coordinates, defining the needed integrals requires more rigor.

Index

Geological Storage of CO_2: Modeling Approaches for Large-Scale Simulation, First Edition.
Jan M. Nordbotten and Michael A. Celia.